Orbital Peril

Jonathon Wetzel

Published by JWetzelBooks, 2025.

ORBITAL PERIL

First edition. April 1, 2025.

ISBN: 979-8227658791

Written by Jonathon Wetzel.

Table of Contents

The Space Junk Crisis: Challenges and Solutions for the Orbital Environment

Introduction

In the vast expanse of Earth's orbit, a crisis has been steadily brewing for decades. Since the dawn of the space age in 1957 with the launch of Sputnik, humanity has been placing objects into orbit at an ever-increasing rate. While these satellites, spacecraft, and stations have revolutionized our world through telecommunications, navigation, weather forecasting, and scientific discovery, they have also created an unintended consequence: space junk.

Space junk, also known as orbital debris, consists of defunct satellites, spent rocket stages, fragments from collisions, and even tools lost during spacewalks. Today, this debris poses a significant threat to operational satellites, space stations, and future space exploration. As of 2023, over 36,000 pieces of debris larger than 10 centimeters are being tracked, along with an estimated 1 million pieces between 1 and 10 centimeters, and countless millions of pieces smaller than 1-centimeter orbiting Earth. Even these tiny fragments, traveling at speeds of approximately 28,000 kilometers per hour—fast enough to circle the Earth in just 90 minutes—can cause catastrophic damage to operational spacecraft.

The growing population of orbital debris has raised concerns among space agencies, commercial space companies, and international organizations. As we continue to launch satellites at unprecedented rates, particularly with the emergence of mega-constellations consisting of thousands of satellites, the risk of collisions increases exponentially. This, in turn, could trigger a cascade of collisions known as the Kessler Syndrome, potentially rendering certain orbital regions unusable for generations.

This book explores the origins, current state, and future projections of the space junk crisis. We will delve into the technical aspects of debris tracking and mitigation, examine the legal and policy frameworks governing space activities, and discuss innovative solutions being developed to address this growing problem. By understanding the complexities of the space junk crisis, we can work toward ensuring the sustainable use of Earth's orbital environment for future generations—preserving humanity's gateway to the stars.

Chapter 1: Understanding Space Junk

Definition and Categorization of Space Junk

Space junk, or orbital debris, encompasses all human-made objects in Earth's orbit that no longer serve a useful function. This includes defunct satellites, spent rocket stages, mission-related debris, and fragments resulting from collisions or explosions. To effectively address the space junk crisis, it is essential to understand the various categories of debris, their origins, and their behaviors in the orbital environment.

Categories of Space Debris

Inactive Payloads: These are satellites that have reached the end of their operational life or have experienced fatal malfunctions. Examples include early communication satellites like Telstar and weather satellites that have long exceeded their design lifetimes. As of 2023, inactive payloads constitute approximately 20% of all trackable objects in orbit, representing billions of dollars of defunct technology silently circling our planet.

Rocket Bodies: These are the upper stages of launch vehicles that remain in orbit after delivering their payloads. Historically, these objects were not deorbited after mission completion, leaving thousands of large rocket bodies in various orbital regions. These objects are particularly concerning due to their size and the potential for uncontrolled reentries. The Russian Proton rocket upper stages and American Centaur upper stages represent significant populations of this debris category, with some measuring the size of a school bus and weighing several tons.

Mission-Related Debris: This category includes objects intentionally released during normal mission operations. Examples include separation bolts, lens covers, deployment mechanisms, and other hardware necessary for spacecraft operation but subsequently discarded. While individual pieces may be small, their cumulative contribution to the debris environment is significant—like a trail of breadcrumbs left behind by each mission.

Fragmentation Debris: This is the most numerous categories of space debris, resulting from explosions or collisions in orbit. Fragmentation events can produce thousands of debris pieces from a single object. Notable events include the 2007 Chinese anti-satellite test against the Fengyun-1C weather satellite, which created over 3,000 trackable fragments, and the 2009 collision between the active Iridium 33 and defunct Cosmos 2251 satellites, which generated more than 2,000 trackable pieces. These violent events effectively create orbital shrapnel fields that persist for decades or centuries.

Deterioration Debris: This category includes paint flakes, thermal blanket fragments, and other materials that detach from spacecraft surfaces due to the harsh space environment. Exposure to atomic oxygen, radiation, micrometeoroids, and thermal cycling causes materials to degrade over time, creating small but potentially damaging debris particles. Even a flake of paint, accelerated to orbital velocities, can damage spacecraft windows or solar panels.

NaK Droplets: A unique category of debris consisting of sodium-potassium (NaK) coolant droplets released from Russian RORSAT nuclear-powered satellites. These spherical droplets, ranging from a few millimeters to centimeters in size, were released when the reactor cores were ejected into higher disposal orbits. Approximately 16 such ejection events occurred, releasing thousands of droplets that remain in orbit today—a legacy of early nuclear power experiments in space.

Size Distribution

The size distribution of orbital debris follows a power law relationship, with the number of objects increasing exponentially as size decreases:

Large Objects (>10 cm): Approximately 36,000 objects in this category are currently tracked by the U.S. Space Surveillance Network and other tracking systems worldwide. These include intact spacecraft, rocket bodies, and large fragments. Objects of this size can cause catastrophic damage to operational satellites and potentially destroy them upon impact—equivalent to a collision with a bowling ball at ten times the speed of a bullet.

Medium Objects (1-10 cm): An estimated 1 million objects fall within this size range. These objects are generally too small to be tracked consistently but large enough to cause significant damage. A 1 cm object striking a satellite can release energy equivalent to the explosion of a hand grenade, potentially disabling critical systems or causing complete mission failure.

Small Objects (<1 cm): Hundreds of millions of debris particles smaller than 1 centimeter orbit Earth. While individually less destructive than larger objects, they can still cause significant damage to spacecraft components, particularly solar arrays, optics, and thinly shielded areas. Their high velocity compensates for their small mass, allowing them to penetrate materials and damage internal systems—much like how a small bullet can pierce armor if traveling fast enough.

Orbital Regions and Debris Distribution

The distribution of space debris is not uniform across all orbital regions. Understanding the concentration of debris in different regions helps in assessing risks and developing targeted mitigation strategies:

Low Earth Orbit (LEO, 200-2,000 km): This region contains the highest concentration of orbital debris, particularly between 750 and 1,000 km altitude. LEO is home to the International Space Station, Earth observation satellites, and many communications constellation satellites. The density of objects in this region poses significant operational challenges and collision risks, creating a congested orbital highway where spacecraft must constantly monitor for potential collisions.

Medium Earth Orbit (MEO, 2,000-35,586 km): This region contains fewer debris objects but includes critical navigation satellite constellations like GPS, GLONASS, and Galileo. The lower debris density is partially offset by the long orbital lifetimes in this region, where objects can remain for thousands of years without atmospheric drag to remove them. This makes MEO particularly vulnerable to long-term accumulation of debris.

Geosynchronous Earth Orbit (GEO, approximately 35,786 km): This valuable orbital region hosts telecommunications and weather satellites. While the debris population is lower than in LEO, the limited nature of the GEO belt makes it a critical resource that must be protected. Objects in GEO have theoretical orbital lifetimes of millions of years—a nearly permanent contamination of this valuable orbital real estate.

Highly Elliptical Orbits (HEO): These orbits, which include Molniya and Tundra orbits, pass through multiple altitude regions, exposing them to varying debris populations. The debris risk for objects in these orbits is complex to assess due to their constantly changing altitude, like a vehicle that must navigate through both crowded city streets and open highways during its journey.

Detection and Tracking Limitations

Our understanding of the space debris environment is constrained by the limitations of current tracking technologies:

Radar Systems: Ground-based radar systems can typically detect objects larger than 10 cm in LEO and 1 meter in GEO. Examples include the U.S. Space Fence, the German TIRA system, and the Russian Okno optical electronic complex. These systems act as traffic cameras for space, monitoring objects as they pass overhead.

Optical Telescopes: These systems are more effective for tracking objects in higher orbits but are limited by weather conditions, daylight operations, and minimum detectable object size (typically 10-30 cm in GEO). Like trying to spot a dark marble against the night sky, optical tracking requires precise instruments and favorable viewing conditions.

In-situ Measurements: Spacecraft like the International Space Station use impact detectors and examine returned surfaces to assess the population of very small debris that cannot be tracked from Earth. These direct observations provide crucial data about the smallest and most numerous debris population.

These limitations create significant uncertainty in our understanding of the total debris population, particularly for objects smaller than 10 cm. This "knowledge gap" complicates risk assessments and mitigation planning—we are trying to solve a problem whose full extent remains partially hidden from our view.

Historical Overview of the First Space Debris Incidents

The history of space debris begins with the first artificial satellite launches and has evolved through a series of significant events that have shaped our understanding of the orbital debris environment. Examining these historical incidents provides valuable context for understanding the current crisis.

The Beginning: 1957-1965

The space age began on October 4, 1957, when the Soviet Union launched Sputnik 1, the first artificial satellite. While Sputnik 1 itself eventually reentered Earth's atmosphere and burned up, its launch vehicle's upper stage became the first piece of space junk. This pattern—operational satellites accompanied by defunct rocket bodies—established the first source of orbital debris.

By 1961, the first recorded fragmentation event occurred when the Ablestar upper stage of the Transit 4-A mission exploded in orbit, creating the first major debris cloud. This explosion, caused by residual propellants in the stage, produced over 300 trackable fragments and established a pattern that would repeat many times over subsequent decades. Between 1957 and 1965, approximately 8,000 objects were placed in orbit, with a significant percentage becoming debris—the earliest signs of an environmental problem that would grow exponentially in the coming decades.

Growing Awareness: 1966-1980

During this period, space agencies began to recognize the potential implications of orbital debris. In 1966, atmospheric scientist John W. Freeman published a paper on the increasing population of human-made objects in space, marking one of the first scientific acknowledgments of the emerging problem.

The 1970s saw several significant developments in space debris awareness. The North American Aerospace Defense Command (NORAD) expanded its tracking capabilities and began maintaining a more comprehensive catalog of orbital objects. Concurrently, NASA scientists led by Donald Kessler began modeling the long-term evolution of the debris environment.

A pivotal moment came in 1978 when NASA's Donald Kessler and Burton Cour-Palais published their landmark paper, "Collision Frequency of Artificial Satellites: The Creation of a Debris Belt." This paper introduced what would later be known as the "Kessler Syndrome," a scenario in which the density of objects in low Earth orbit becomes high enough that collisions between objects could cause a cascade effect, with each collision generating debris that increases the likelihood of further collisions—a space version of a chain reaction accident on a highway.

This period also saw the first deliberate anti-satellite (ASAT) tests, with the Soviet Union conducting several tests of their co-orbital ASAT system between 1968 and 1982. While these tests did create some debris, they were designed to minimize fragmentation compared to later ASAT demonstrations.

First Major Incidents: 1981-2000

The 1980s and 1990s witnessed several events that elevated awareness of the space debris issue:

Cosmos 1275 Fragmentation (1981): This Soviet navigation satellite spontaneously exploded in orbit approximately five months after launch, creating over 300 trackable debris pieces. The cause was likely a battery explosion, highlighting the risk posed by stored energy sources on inactive satellites.

Solar Maximum Mission Examination (1984): When the Solar Maximum Mission satellite was retrieved by the Space Shuttle Challenger for repair, engineers discovered more than 1,000 microcrater impacts on its surface. This provided the first direct evidence of the prevalence of small debris particles that couldn't be tracked from the ground—a sobering revelation about the true extent of the orbital debris environment.

First Confirmed Collision (1991): The Russian Cosmos 1934 satellite was struck by debris from another Russian satellite, Cosmos 926. This event marked the first confirmed accidental collision between cataloged objects in orbit, though the collision did not create significant additional debris.

First Hypervelocity Impact on an Operational Satellite (1996): A French microsatellite, Cerise, was struck by a cataloged fragment from an Ariane rocket upper stage that had exploded a decade earlier. The impact severed Cerise's gravity gradient stabilization boom but did not disable the satellite completely. This incident marked the first documented case of a functional satellite being damaged by a tracked piece of debris—a warning of the vulnerability of operational spacecraft.

Space Shuttle Window Replacements: Throughout the 1990s, NASA had to replace numerous Space Shuttle windows that had been damaged by impacts with paint flecks and other small debris. Each window replacement cost approximately $50,000, highlighting the economic impact of even tiny debris particles and demonstrating that no spacecraft, not even crewed vehicles, were immune to the debris problem.

By the end of this period, the total number of cataloged objects had grown to over 8,500, with approximately 4,500 remaining in orbit. Space agencies began implementing the first debris mitigation guidelines, including passivation (removing stored energy) of upper stages and designing satellites with end-of-life disposal capabilities.

Modern Debris Crisis Catalysts: 2001-Present

The contemporary space debris crisis was significantly exacerbated by several high-profile events in the 21st century:

Chinese ASAT Test (2007): In January 2007, China conducted an anti-satellite missile test, destroying their defunct Fengyun-1C weather satellite. This single event created over 3,500 trackable debris fragments (and an estimated 150,000 particles larger than 1 cm), instantly becoming the largest debris-generating event in history. The test was particularly problematic because it occurred at a high altitude (approximately 865 km), where atmospheric drag is minimal, meaning the debris will remain in orbit for decades or even centuries—a persistent cloud of orbital shrapnel threatening all spacecraft in its path.

Iridium-Cosmos Collision (2009): On February 10, 2009, an active commercial communications satellite, Iridium 33, collided with a defunct Russian military satellite, Cosmos 2251. The collision occurred at an altitude of approximately 789 km and a relative velocity of about 11.7 km/s, creating more than 2,300 trackable debris fragments. This event was significant as it represented the first major accidental collision between two intact satellites and validated concerns about the increasing collision risk in popular orbital regimes—a real-world demonstration of the Kessler Syndrome theory in action.

Indian ASAT Test (2019): India conducted its own anti-satellite test, Mission Shakti, targeting the Indian Microsat-R satellite at a lower altitude of approximately 300 km. While the test created several hundred trackable fragments, the lower altitude meant most debris reentered Earth's atmosphere within weeks or months, mitigating the long-term impact compared to the Chinese test—showing that if such tests must be conducted, altitude selection makes a critical difference.

Russian ASAT Test (2021): Russia destroyed its Cosmos 1408 satellite with a direct-ascent anti-satellite missile, creating over 1,500 trackable fragments at an altitude that posed direct threats to the International Space Station and other human spaceflight activities. The crew of the ISS had to take shelter in their spacecraft as a precautionary measure as debris clouds passed near the station—a stark reminder that space debris isn't just a technical problem but a direct threat to human lives in space.

These major events, combined with the continuous launch of new satellites and the fragmentation of existing space objects, have accelerated the growth of the orbital debris population. The emergence of mega-constellations, such as SpaceX's Starlink and OneWeb, which plan to deploy thousands of satellites, has further complicated the space traffic management challenge.

Impact of Early Debris Events on Policy and Technology

The early space debris incidents catalyzed important developments in both policy and technology:

Establishment of the Inter-Agency Space Debris Coordination Committee (IADC) in 1993, bringing together space agencies from around the world to address the debris issue collectively.

Development of the first international debris mitigation guidelines by the IADC in 2002, which were later adopted by the United Nations Committee on the Peaceful Uses of Outer Space (COPUOS).

Creation of specialized debris tracking networks and improvements to existing systems to enhance detection capabilities for smaller objects.

Research into spacecraft shielding technologies, such as the Whipple shield used on the International Space Station, designed to protect against hypervelocity impacts from small debris particles.

Implementation of collision avoidance procedures for operational satellites, with the first documented collision avoidance maneuver performed by a commercial satellite operator in the late 1990s.

These historical events and responses have shaped our current understanding of the space debris environment and informed the development of mitigation and remediation strategies that will be explored in subsequent chapters. The lessons learned from these incidents highlight the need for international cooperation, technological innovation, and regulatory frameworks to address the growing challenge of space debris—ensuring that our journey to the stars doesn't end at the edge of our own atmosphere.

Chapter 2: The Kessler Syndrome

The Orbital Chain Reaction That Threatens Space Access

In 1978, NASA scientist Donald J. Kessler described a scenario that now haunts space agencies and satellite operators worldwide. He envisioned a future where collisions between orbiting objects create debris that triggers more collisions, eventually rendering certain orbital regions unusable for generations. This theoretical scenario, now known as the "Kessler Syndrome," represents one of the most concerning potential outcomes of our growing space debris crisis.

The Fundamental Concept: A Self-Sustaining Cascade

At its core, the Kessler Syndrome describes a tipping point in orbital congestion. When the density of objects in a particular orbital region reaches a critical threshold, the probability of collisions increases significantly. Each collision generates additional debris fragments, which in turn increase the likelihood of subsequent collisions. This creates a positive feedback loop—a chain reaction where debris generation outpaces natural removal processes such as atmospheric drag.

What makes the Kessler Syndrome particularly troubling is its self-perpetuating nature. Unlike most terrestrial pollution problems, where stopping the polluting activity allows the environment to recover, a debris cascade continues to worsen even if we launch nothing new into orbit. Once triggered, this process could continue for decades or centuries, potentially closing off valuable orbital regions to future generations.

The Physics of Orbital Destruction

To understand why the Kessler Syndrome poses such a significant threat, we need to examine the extreme physics of orbital collisions:

Hypervelocity Impacts: In Low Earth Orbit (LEO), objects typically travel at approximately 7.5 kilometers per second—more than 27,000 kilometers per hour. When two objects collide, their relative velocity can exceed 10 kilometers per second, more than ten times faster than a bullet. At these incredible speeds, collisions don't simply result in objects bouncing off each other; they cause catastrophic fragmentation, with materials behaving more like liquids upon impact. A seemingly insignificant 1-centimeter object striking a satellite can release energy equivalent to a hand grenade explosion.

Exponential Fragment Generation: A single collision between two intact satellites can generate thousands of trackable fragments and millions of smaller particles. The 2009 collision between the operational Iridium 33 communications satellite and the defunct Cosmos 2251 satellite demonstrates this destructive potential—it produced over 2,300 trackable fragments larger than 10 centimeters and an estimated 35,000 fragments larger than 1 centimeter. Each of these fragments becomes a potential impactor for other objects in orbit.

Orbital Persistence: Debris generated at higher altitudes can remain in orbit for decades, centuries, or even millennia. At altitudes above 800 kilometers, atmospheric drag becomes minimal, providing little natural cleansing of the orbital environment. Fragments from the 2007 Chinese anti-satellite weapon test conducted at 865 kilometers altitude are expected to remain in orbit for more than a century, continuing to threaten operational satellites throughout that period.

Orbital Crossing Points: Due to the nature of orbital mechanics, certain inclinations and altitudes become congested with crossing orbital planes. These crossing points create high-risk areas for potential collisions. Popular orbital regimes, such as sun-synchronous orbits used for Earth observation satellites, experience higher traffic and consequently higher collision risks.

When Does the Cascade Begin?

The Kessler Syndrome doesn't occur instantaneously but represents a transition from a stable environment to an unstable one. This transition depends on several key factors:

Critical Density Threshold: The cascade begins when the spatial density of objects exceeds a critical value—the point where the rate of new debris generation through collisions exceeds the rate at which debris is removed by natural processes. While this threshold varies by orbital altitude, Kessler's original calculations suggested that some regions of LEO were approaching this critical density by the late 1970s, a situation that has only worsened in subsequent decades.

Runaway Timescale: Even after crossing the critical threshold, the full cascade develops over time, not instantly. Kessler's models suggested timescales of decades to centuries for the cascade to fully manifest, depending on the initial conditions and the rate of new launches. Recent studies with updated object catalogs and improved collision models suggest that in some heavily congested orbital regions, this process may already be underway.

Altitude Dependence: The risk of a cascade is highest in regions where debris persistence is longest and object density is highest. In LEO, this typically corresponds to altitudes between 750 and 1,000 kilometers, where many sun-synchronous and communication satellites operate, and atmospheric drag is minimal compared to lower altitudes.

Mass Distribution Factors: The total mass in orbit and how that mass is distributed among objects affects cascade development. A few large objects present less collision risk than many smaller objects of the same total mass. However, when large objects do collide, they generate many more fragments, potentially accelerating the cascade.

Warning Signs: Are We Already on the Path?

Several events and observations in recent decades suggest we may be approaching Kessler Syndrome conditions in certain orbital regions:

Increasing Collision Avoidance Maneuvers: The International Space Station (ISS) performed its first debris avoidance maneuver in 1999. By 2020, such maneuvers had become routine, with the ISS executing multiple avoidance maneuvers annually. Commercial satellite operators now perform thousands of collision avoidance maneuvers each year, with the number growing steadily.

Rising Background Flux: In-situ measurements from returned spacecraft surfaces show an increasing flux of small particles in LEO. The Long Duration Exposure Facility (LDEF), which spent 5.7 years in orbit from 1984 to 1990, documented impact cratering rates that exceeded predictions based solely on the natural micrometeoroid environment, indicating a growing population of small human-made debris.

Collisions Between Cataloged Objects: The 2009 Iridium-Cosmos collision demonstrated that even with tracking capabilities, high-energy collisions between large objects can and do occur. Subsequent analysis revealed that dozens of similar close approaches occur daily, with current tracking and prediction capabilities insufficient to warn of all potential collisions.

Fragmentation Events: Between 2007 and 2022, over 20 major fragmentation events created thousands of new trackable debris objects. Some of these were deliberate (like anti-satellite tests), while others resulted from battery explosions, propellant tank ruptures, or unknown causes affecting defunct satellites and rocket bodies.

Mega-Constellation Deployments: The rapid increase in satellite numbers due to mega-constellations like SpaceX's Starlink (which alone plans to deploy more than 12,000 satellites) represents an unprecedented growth in the orbital population. Even with high reliability and responsible end-of-life disposal, the sheer number of objects increases the statistical probability of failures that could contribute to the debris environment.

The Day After Tomorrow: Life in a Post-Kessler World

If the Kessler Syndrome were to fully manifest in certain orbital regions, the consequences would be far-reaching:

Operational Hazards: Satellites in affected orbital regions would face significantly higher collision risks, potentially requiring more frequent replacements and more robust shielding, increasing costs and reducing operational lifetimes.

Access Limitations: Certain orbital regions could become effectively unusable for decades or centuries. This would be particularly problematic for unique orbital regimes that serve specific purposes, such as sun-synchronous orbits for Earth observation or geosynchronous orbit for telecommunications.

Transit Risks: Even if operations were to focus on unaffected orbital regions, spacecraft would still need to transit through debris-congested regions during launch and orbit raising, increasing mission risks.

Economic Impacts: The global space economy, valued at over $400 billion annually as of 2023, relies heavily on satellite infrastructure. Degradation of the orbital environment would have significant economic repercussions for telecommunications, Earth observation, navigation, and other space-dependent services.

Scientific Setbacks: Space-based astronomy and Earth science missions could be compromised, limiting our ability to study climate change, perform astronomical observations, and monitor global environmental conditions.

Long Recovery Timeline: Unlike terrestrial environmental problems, orbital debris at higher altitudes remains in place for extremely long periods. Once a full cascade begins, recovery could take centuries even with active debris removal interventions.

Looking Through the Mathematical Lens

Understanding the Kessler Syndrome requires sophisticated mathematical models and computer simulations to predict collision probabilities, debris generation rates, and long-term environmental evolution. These models have evolved significantly since Kessler's original work in the 1970s, incorporating more detailed physical processes and utilizing increasingly comprehensive data on the orbital debris population.

Fundamental Equations and Parameters

At the core of debris environment modeling are several key equations and parameters:

Spatial Density Equation: The spatial density of objects in a given orbital region is typically expressed as:

$\rho(h,i,\Omega) = N / V$

Where:

ρ is the spatial density (objects per cubic kilometer)

N is the number of objects in the volume

V is the volume of the orbital shell

h, i, and Ω represent altitude, inclination, and right ascension of ascending node

Collision Probability Calculation: For any two objects, the probability of collision during a close approach can be calculated using:

$P\_c = (\sigma \times v\_rel \times \Delta t) / V\_encounter$

Where:

P_c is the collision probability

σ is the combined cross-sectional area of the objects

v_rel is the relative velocity

Δt is the time duration of the encounter

V_encounter is the volume of the encounter region

Debris Generation Models: The number of fragments generated in a collision can be estimated using the NASA Standard Breakup Model:

$N(L\_c) = 0.1 \times M\text{\^{}}0.75 \times L\_c\text{\^{}}\text{-}1.71$

Where:

N(L_c) is the number of fragments larger than characteristic length L_c

M is the mass of the target in kilograms

L_c is the characteristic length in meters

Orbital Lifetime Calculation: The time required for an object to decay from its initial orbit due to atmospheric drag can be estimated using:

$$\tau = (m / C_D A) \times (H / \rho_0) \times e^{(h/H)}$$

Where:

τ is the orbital lifetime

m is the object mass

C_D is the drag coefficient

A is the cross-sectional area

H is the atmospheric scale height

ρ_0 is the reference atmospheric density

h is the initial altitude

These fundamental equations form the basis for more complex numerical simulations that model the long-term evolution of the debris environment.

Evolution of Debris Environment Models

Debris modeling has advanced significantly since Kessler's early work:

First-Generation Models (1970s-1980s): Kessler's original model used simplified orbital mechanics and basic fragmentation assumptions. These models treated orbital regions as shells with uniform debris distributions and used statistical methods to estimate collision probabilities. Despite their simplicity, these models correctly predicted the general trends we now observe in the orbital environment.

Second-Generation Models (1990s-2000s): NASA's EVOLVE model and similar tools incorporated more realistic orbital mechanics, including perturbation effects from Earth's non-spherical gravity field, solar radiation pressure, and atmospheric drag. These models could track individual objects rather than just statistical populations and included improved breakup models based on laboratory hypervelocity impact tests and observed on-orbit fragmentations.

Current-Generation Models (2010s-Present): Modern debris models like NASA's LEGEND (LEO-to-GEO Environment Debris model), ESA's MASTER (Meteoroid and Space Debris Terrestrial Environment Reference), and the multi-agency ORDEM (Orbital Debris Engineering Model) incorporate:

High-fidelity orbital propagators that account for all significant perturbation forces

Detailed object catalogs with specific characteristics for each tracked object

Improved breakup models based on observed fragmentation events

Future launch traffic projections based on historical patterns and announced plans

Mitigation measure effectiveness assessments

Monte Carlo simulation techniques to account for uncertainties

These models typically run thousands of simulations with varying initial conditions to develop probability distributions of possible future scenarios.

What the Numbers Tell Us

Modern debris environment simulations have yielded several important insights:

Instability Thresholds: Updated simulations confirm Kessler's fundamental hypothesis about critical density thresholds but offer more precise estimates of when and where these thresholds may be reached. Current models suggest that the 850-1000 km altitude range may already be above the critical threshold where debris generation exceeds natural removal processes, even without new launches.

Collision Rate Projections: Without additional mitigation measures, the expected collision rate between objects larger than 10 cm in LEO could increase from approximately one catastrophic collision every 5-10 years (current rate) to one every 3-5 years by 2050. Over centuries, this rate could further increase to multiple collisions per year if the debris population continues to grow.

Debris Source Analysis: Modern simulations can differentiate between various sources of debris generation. Recent analyses suggest that while explosions of rocket bodies and satellites were historically the dominant source of debris, collisions are likely to become the primary source in the coming decades if the current trajectory continues.

Mitigation Effectiveness: Simulations evaluating different mitigation strategies indicate that:

Post-mission disposal (deorbiting) of satellites and rocket stages is effective at limiting long-term population growth

A 90% compliance rate with the 25-year rule (removing objects from LEO within 25 years of mission completion) significantly reduces long-term debris accumulation

Even with perfect compliance with current mitigation guidelines, active debris removal may be necessary to stabilize some orbital regions

Long-Term Projections: Extended simulations over 100-200 year timeframes suggest that without intervention, certain orbital regions could experience exponential growth in the debris population, potentially reaching densities 10-20 times higher than present levels by the end of the century.

Case Study: Four Futures for Low Earth Orbit

A comprehensive modeling study conducted by an international research team in 2021 examined the potential evolution of the LEO debris environment under different scenarios. The study utilized a high-fidelity orbital propagator coupled with a detailed breakup model and considered four scenarios:

Business as Usual (BAU): Continuation of current launch rates with existing compliance levels for post-mission disposal (approximately 60% compliance with the 25-year rule).

Improved Mitigation (IM): Increased compliance with post-mission disposal (90% compliance with the 25-year rule) and passivation of all rocket bodies to prevent explosions.

Active Debris Removal (ADR): Implementation of a program to actively remove 5-10 large, defunct objects from high-risk orbits annually, in addition to improved mitigation measures.

Technological Breakthrough (TB): Development and deployment of technologies that can achieve nearly 100% successful post-mission disposal plus active removal of 20 objects annually.

The simulation was run 100 times for each scenario with Monte Carlo variations to account for the probabilistic nature of collisions. The results were striking:

In the BAU scenario, the number of catastrophic collisions averaged 28 over the 100-year simulation period, leading to a nearly 300% increase in the total debris population.

The IM scenario reduced collisions to an average of 12 over the century, with a debris population increase of approximately 110%.

The ADR scenario further reduced collisions to an average of 5, with a debris population increase of only about 30%.

Only the TB scenario achieved stability, with the debris population actually decreasing by approximately 20% over the century.

This case study demonstrates both the severity of the potential cascade effect and the importance of combining multiple approaches to address the problem effectively.

The Crystal Ball is Cloudy: Limitations and Uncertainties

Despite their sophistication, current debris environment models face several significant limitations:

Incomplete Observational Data: Current tracking capabilities cannot consistently detect objects smaller than 10 cm in LEO or 1 meter in GEO, creating significant uncertainty about the true population of smaller debris objects that can still cause significant damage.

Breakup Model Uncertainties: The physical processes of satellite fragmentation in hypervelocity impacts are not perfectly understood, particularly for modern composite materials, leading to uncertainties in fragment generation predictions.

Future Launch Traffic: Long-term projections require assumptions about future launch rates, satellite sizes, and constellation deployments that may diverge significantly from actual developments.

Mitigation Compliance: The effectiveness of debris mitigation guidelines depends on compliance rates, which can vary by country, organization, and market conditions.

Solar Activity Effects: Long-term atmospheric drag depends on solar activity, which affects atmospheric density at orbital altitudes. Solar cycle predictions beyond a few years carry significant uncertainty, affecting orbital lifetime calculations.

Computational Limitations: The most detailed simulations require significant computational resources, often forcing trade-offs between fidelity, time horizon, and the number of Monte Carlo iterations.

These limitations mean that while models provide valuable insights into trends and relative comparisons between scenarios, precise predictions of absolute debris population values or collision dates should be interpreted with appropriate caution.

Future Directions in Debris Modeling

Several advances in debris modeling are currently being pursued to address these limitations:

Improved Observation Integration: New models are being developed to better incorporate data from optical, radar, and in-situ sensors to create more accurate debris population estimates, particularly for smaller size ranges.

Machine Learning Approaches: Artificial intelligence techniques are being applied to identify patterns in collision probabilities and improve predictive capabilities beyond what traditional statistical methods can achieve.

High-Performance Computing: Leveraging supercomputing resources allows for higher-fidelity simulations with more Monte Carlo iterations, reducing statistical uncertainties.

Economic Integration: Newer models are beginning to incorporate economic factors, linking debris mitigation costs with the economic consequences of collisions to better inform policy decisions.

Unified International Modeling: Efforts to standardize assumptions across different national and organizational models are underway to create more consistent global assessments of the debris environment.

These advancements will continue to improve our understanding of the Kessler Syndrome and help guide effective response strategies.

As we stand at this critical juncture in space development, the decisions we make in the coming years regarding debris mitigation, active removal technologies, and international cooperation will determine whether the Kessler Syndrome remains a theoretical concern or becomes an unfortunate reality that constrains humanity's access to space for generations to come.

Chapter 3: A Historical Perspective

Milestones in Space Exploration and Space Junk Generation

The accumulation of space debris parallels the history of space exploration itself. Each major advancement in human space capabilities has introduced more objects into Earth's orbit, contributing to the growing challenge we face today. By examining this historical progression, we can better understand the current crisis and identify patterns that may inform future solutions.

The Dawn of the Space Age (1957-1965)

Humanity's venture into space began with the Soviet Union's launch of Sputnik 1 on October 4, 1957. This 83.6-kilogram satellite orbited Earth for three months before its orbit decayed. While Sputnik itself didn't become long-lasting debris, its rocket body remained in orbit significantly longer—establishing a pattern that would define the space age: operational satellites accompanied by non-functional rocket components.

Several significant debris-generating events marked these early years:

Vanguard 1 (1958): Launched by the United States as the fourth artificial satellite in orbit, Vanguard 1 now holds the distinction of being the oldest human-made object still in space. With an expected orbital lifetime exceeding 200 years, it represents one of our first permanent contributions to the orbital environment.

Project West Ford (1961-1963): In a controversial experiment, the U.S. military released 480 million copper dipole antennas into medium Earth orbit to test creating an artificial ionosphere for global communications. Though many of these tiny copper needles re-entered Earth's atmosphere within a few years, some clusters remain in orbit today—marking one of the first instances of intentionally creating numerous artificial objects in space.

First Accidental Explosion (1961): An Ablestar upper stage from the Transit 4-A mission exploded in orbit due to residual propellants, creating the first significant debris cloud from a single event. This fragmentation produced hundreds of trackable objects and established spent rocket stage explosions as a major debris source.

Early Anti-Satellite Testing: Both the United States (Program 437) and Soviet Union began developing anti-satellite weapons during this period, though the early tests generally limited debris generation.

By 1965, approximately 700 objects were being tracked in Earth orbit. The primary debris sources during this pioneering era were spent rocket stages, mission-related objects (lens covers, separation mechanisms, etc.), and a few fragmentation events.

The Race to the Moon and Early Space Stations (1966-1979)

This period witnessed both the culmination of the Moon race and the first long-duration human presence in Earth orbit through space stations. Key developments included:

Apollo Program (1968-1972): The Apollo missions to the Moon left debris in both Earth orbit and lunar orbit. Some Apollo components, such as the third stage of Apollo 12's Saturn V rocket, remain in heliocentric orbit today—human-made artifacts traveling through the solar system.

Salyut and Skylab Space Stations (1971-1979): These first-generation space stations represented larger structures in orbit. Skylab's uncontrolled reentry in 1979 captured international attention, high-

lighting the challenges of managing large objects at the end of their operational lives. While most of Skylab burned up during reentry or fell into the ocean, its descent demonstrated the complexity of spacecraft disposal.

Growth in Geosynchronous Orbit Usage: The 1970s saw increased utilization of geosynchronous orbit for communications satellites. Unlike objects in lower orbits, satellites at this altitude (approximately 36,000 kilometers) remain in space indefinitely unless actively moved to disposal orbits—creating a permanent legacy of human activity.

First Awareness of the Debris Problem: In 1978, NASA scientists Donald Kessler and Burton Cour-Palais published their seminal paper on the cascade effect (later known as the "Kessler Syndrome"), bringing scientific attention to the long-term implications of space debris. This marked the beginning of formal debris research programs at NASA and other space agencies.

Cosmos 954 Reentry (1978): This Soviet nuclear-powered satellite made an uncontrolled reentry over Canada, spreading radioactive debris across a wide area. While not directly related to orbital debris, this incident highlighted the potential hazards of uncontrolled spacecraft reentries and spurred international discussion about space environmental responsibilities.

By 1980, the tracked space object catalog had grown to approximately 4,700 objects, with over 75% classified as debris rather than operational payloads. Importantly, this period saw the first recognition that space debris represented a long-term environmental challenge requiring dedicated research and potential regulation.

Commercialization and Expanded Space Activities (1980-1999)

The 1980s and 1990s witnessed significant expansion and diversification of space activities, including commercial services, large-scale scientific missions, and the beginning of the International Space Station program. Notable developments included:

Space Shuttle Program (1981-2011): The Space Shuttle generated relatively little long-term debris compared to expendable launch vehicles, as its main components were either reused or designed to reenter Earth's atmosphere. However, Shuttle missions occasionally released small items during spacewalks and deployments.

Debris from ASAT Tests: In September 1985, the United States conducted an anti-satellite test against the Solwind P78-1 satellite, creating approximately 285 trackable fragments. Despite being conducted at a relatively low altitude (525 km), some fragments remained in orbit for over a decade, demonstrating the persistence of even "short-lived" debris.

First Collision Avoidance Maneuvers: By the late 1980s, NASA developed formal collision avoidance procedures for the Space Shuttle based on tracking data from the U.S. Space Surveillance Network. The first documented collision avoidance maneuver for the Shuttle occurred in 1991—an operational measure that would become increasingly common in later decades.

Proliferation of Upper Stage Explosions: Throughout this period, numerous rocket upper stages experienced explosions in orbit due to residual propellants or battery failures. The Delta second stage, Ariane third stage, and various Soviet/Russian stages were particularly problematic, with over 40 fragmentation events recorded between 1980 and 1999.

Mir Space Station Operations (1986-2001): The Russian Mir space station represented a long-duration human presence in LEO and ex-

perienced several notable incidents, including a collision with a Progress supply vehicle in 1997. While not debris-related, this collision demonstrated the potential consequences of orbital impacts.

Early Mitigation Guidelines: By the late 1990s, space agencies began developing the first formal debris mitigation guidelines. NASA issued its first guidelines in 1995, followed by similar documents from other major spacefaring nations—marking the transition from awareness to active policy development.

International Cooperation on Debris: The Inter-Agency Space Debris Coordination Committee (IADC) was established in 1993 as the first international forum specifically focused on the orbital debris issue, marking an important step toward global cooperation.

By the end of this period, the tracked catalog had grown to approximately 8,500 objects, with models suggesting hundreds of thousands of additional untracked smaller objects in orbit. The increasing value of space assets and growing recognition of collision risks led to more serious consideration of debris mitigation approaches.

The Modern Debris Era (2000-Present)

The past two decades have seen both dramatic increases in the debris population and heightened awareness of space sustainability issues:

Chinese ASAT Test (2007): China's destruction of the Fengyun-1C weather satellite created over 3,500 trackable debris fragments, representing the largest single debris-generating event in history. The high altitude of this test (approximately 865 km) ensured that much of this debris will remain in orbit for decades, significantly elevating collision risks in sun-synchronous orbit.

Iridium-Cosmos Collision (2009): The accidental collision between the operational Iridium 33 satellite and defunct Cosmos 2251 satel-

lite became the first major collision between intact satellites. This event generated over 2,300 trackable fragments and highlighted that such collisions were no longer theoretical concerns but practical realities that could affect operational systems.

Proliferation of CubeSats: Beginning around 2010, standardized small satellites (particularly CubeSats) began to be deployed in large numbers. While individual CubeSats pose minimal debris risk, their growing numbers, sometimes limited tracking capabilities, and occasionally lower reliability have raised concerns about increased collision potential.

Mega-Constellation Era: Starting with SpaceX's Starlink in 2019, satellite operators began deploying unprecedented numbers of satellites in large constellations. SpaceX alone has launched over 4,000 Starlink satellites as of 2023, with plans for at least 12,000. Other companies including OneWeb, Amazon's Project Kuiper, and several Chinese entities have announced plans for constellations ranging from hundreds to thousands of satellites—potentially transforming the orbital landscape.

Additional ASAT Tests: India conducted an ASAT test in 2019 (Mission Shakti), though at a lower altitude that limited long-term debris generation. Russia conducted a widely condemned ASAT test in 2021, destroying the Cosmos 1408 satellite and creating over 1,500 trackable fragments in orbits that threatened the International Space Station.

International Policy Development: The UN Committee on the Peaceful Uses of Outer Space (COPUOS) adopted 21 voluntary guidelines for space sustainability in 2018, and commercial entities have begun developing additional best practices through organizations like the Space Safety Coalition and the World Economic Forum's Space Sustainability Rating.

Experimental Debris Removal: The first practical demonstrations of debris removal technologies have emerged in this period, including JAXA's small-scale experiments, the RemoveDEBRIS mission, and commercial developments by companies like Astroscale and ClearSpace. These missions have tested various capture mechanisms, de-orbiting technologies, and proximity operations necessary for future large-scale debris removal.

As of 2023, the tracked catalog exceeds 36,000 objects, with more than 50% added in just the last decade. Modeling suggests the presence of over 1 million objects larger than 1 cm and hundreds of millions of smaller particles. The current era is characterized by both rapidly increasing utilization of Earth orbit and growing awareness of the need for sustainable space operations.

Analysis of the Impact of Key Events on the Space Junk Population

Certain historical events have had disproportionate impacts on the orbital debris environment. Analyzing these events provides insights into debris generation mechanisms, the effectiveness of mitigation approaches, and potential future risks.

The Chinese ASAT Test (2007): A Case Study in Long-lasting Debris Generation

On January 11, 2007, China conducted an anti-satellite weapon test, using a ground-based missile to destroy their defunct Fengyun-1C weather satellite. This single event fundamentally altered the orbital debris environment:

Unprecedented Debris Generation: The test created approximately 3,500 trackable objects (larger than 10 cm) and an estimated 150,000 debris fragments larger than 1 cm—representing a nearly 25% increase in the total tracked debris population from a single event.

Altitude Implications: Conducted at approximately 865 kilometers, an altitude where atmospheric drag is minimal, modeling indicates that approximately 30% of the debris generated will still be in orbit 100 years after the test, creating a multi-generational legacy.

Orbital Distribution: The polar orbit of Fengyun-1C resulted in debris being spread across a wide range of inclinations, effectively "contaminating" much of the sun-synchronous orbital regime—particularly valuable for Earth observation missions.

Operational Consequences: In the years following the test, satellite operators worldwide reported increased conjunction warnings and collision avoidance maneuvers due to Fengyun debris. The International Space Station has performed multiple debris avoidance maneuvers specifically to avoid fragments from this test.

Policy Impacts: International condemnation of the test helped accelerate discussions about responsible behavior in space and contributed to the development of transparency and confidence-building measures. The event catalyzed subsequent international discussions on space sustainability.

Long-term tracking of the Fengyun-1C debris cloud has provided valuable data on debris dispersion patterns, orbital decay rates, and the accuracy of fragmentation models. This information has improved our ability to predict the consequences of future fragmentation events and design more effective mitigation strategies.

The Iridium-Cosmos Collision (2009): When Probability Becomes Reality

The February 10, 2009 collision between Iridium 33 and Cosmos 2251 marked a turning point in space operations:

Accident Despite Tracking: Unlike the intentional ASAT test, this collision occurred despite both objects being tracked by the U.S. Space Surveillance Network. The event highlighted limitations in conjunction prediction accuracy, tracking data sharing, and collision avoidance protocols.

Operational vs. Defunct Object: The collision involved an operational commercial satellite and a defunct military satellite, raising questions about responsibility for avoidance maneuvers and the growing risk posed by non-maneuverable derelicts.

Economic Consequences: The destruction of the operational Iridium satellite caused direct financial loss and service interruption, providing a concrete example of the economic impacts of orbital debris collisions.

Debris Cloud Characteristics: The collision occurred at a nearly perpendicular angle between the two satellites' orbital planes, resulting in debris being widely dispersed. While some fragments were ejected into short-lived orbits, others were boosted into higher, longer-lasting orbits.

Operational Changes: Following this collision, commercial satellite operators significantly enhanced their collision avoidance procedures. Many companies began subscribing to commercial space situational awareness services to supplement government-provided conjunction warnings, and lowered their collision risk thresholds for performing avoidance maneuvers.

Analysis of this collision has informed improvements in conjunction assessment methods, highlighted the importance of maintaining accurate orbital data on defunct objects, and demonstrated the practical value of satellite maneuverability for collision avoidance.

The Mega-Constellation Era: Quantitative and Qualitative Changes

The deployment of satellite mega-constellations beginning in 2019 represents both a quantitative leap in orbiting objects and qualitative changes in space traffic management:

Scale of Deployment: SpaceX's Starlink constellation alone has added over 4,000 satellites to orbit in just four years—more than the total number of active satellites that existed before the constellation began deployment. When combined with OneWeb, Amazon's Project Kuiper, and various Chinese constellation plans, these systems could add over 20,000 satellites to LEO within the next decade.

Collision Risk Profile: Constellation satellites typically have:

Similar orbital altitudes, creating "shells" of increased object density

Automated collision avoidance capabilities

Designed-in end-of-life disposal features

These characteristics create a mixed risk profile: increased object density but potentially better collision avoidance capabilities than traditional satellites.

Failure Modes and Statistics: With thousands of satellites, even a small failure rate can create significant debris concerns. If 1% of constellation satellites experience propulsion failures preventing controlled deorbit, hundreds of large derelict objects could remain in valuable orbital regions for years.

Astronomical Impacts: While not directly related to collision risks, mega-constellations have raised concerns about light pollution affecting astronomical observations. This has highlighted the broader environmental impacts of large-scale space activities and led to design changes such as DarkSat and VisorSat to reduce satellite brightness.

Regulatory Challenges: Existing regulatory frameworks were not designed for the rapid deployment of thousands of similar satellites. This has prompted reviews of licensing procedures, deorbit requirements, and collision avoidance protocols by the FCC and other national regulators.

Early operational data from mega-constellations shows mixed results for space sustainability. On the positive side, Starlink satellites have performed thousands of autonomous collision avoidance maneuvers and demonstrated high post-mission disposal success rates. However, there have been occasional failures, unexpected behaviors, and conjunction events that have required intervention. The long-term impact of these constellations on the orbital environment remains an area of active study and concern.

Historical Trends in Debris Generation Rates

Analyzing debris generation rates over time reveals several important patterns:

Explosions vs. Collisions: Historically, the majority of fragmentation debris (approximately 70%) has come from explosions rather than collisions. However, this balance is expected to shift in coming decades as collision-generated debris becomes more prevalent, potentially accelerating the Kessler Syndrome.

Mission-Related Debris Reduction: There has been a notable decrease in mission-related debris (lens covers, adapter rings, separation bolts, etc.) since the 1990s, demonstrating the effectiveness of design changes and operational best practices in reducing some debris sources.

Impact of Mitigation Guidelines: The implementation of passivation requirements (removing stored energy from defunct satellites and rocket stages) has reduced the rate of accidental explosions since the mid-2000s, though compliance remains inconsistent across different operators and countries.

Correlation with Launch Activity: While there is some correlation between launch rates and debris generation, the relationship is not linear due to improvements in vehicle design and operations. Modern launch vehicles typically release fewer debris objects than their predecessors, despite often carrying more payloads.

Low vs. High Altitude Trends: Debris at lower altitudes (below 600 km) shows clearer signs of natural cleanup through atmospheric drag, with population levels that rise after major fragmentation events but then gradually decline. At higher altitudes, debris accumulation has been more consistently upward, reflecting the near-permanent nature of high-altitude debris.

These historical trends inform predictions about future debris population growth and help identify the most effective targets for mitigation and remediation efforts.

Technological Evolution and Its Impact on Debris

Changes in spacecraft and launch vehicle technology have significantly influenced debris generation patterns:

Rocket Body Design: Early rocket upper stages frequently experienced on-orbit explosions due to residual propellants and battery failures. Modern designs increasingly incorporate passivation features (propellant depletion burns, battery discharge systems) that have reduced but not eliminated this risk.

Satellite Miniaturization: The trend toward smaller satellites has mixed implications for debris:

Smaller satellites create less debris if they fragment

However, their typically lower cost and higher numbers may lead to more objects in orbit overall

CubeSats and other small satellites often have limited or no propulsion, reducing their ability to perform collision avoidance

Materials Evolution: Modern spacecraft increasingly use composite materials rather than metals for structures. This affects fragmentation patterns when collisions occur, typically resulting in more numerous but lighter debris pieces. Testing at NASA's Hypervelocity Impact Technology facility has shown that carbon fiber composites tend to produce more fragments than aluminum structures of similar mass.

End-of-Life Technologies: Technological developments specifically aimed at reducing debris include:

Drag enhancement devices like sails and balloons to accelerate orbital decay

Tethers for electrodynamic drag or momentum exchange

Specialized deorbit propulsion modules

"Design for demise" approaches that ensure more complete burnup during reentry

Autonomous Systems: The increasing use of autonomous collision avoidance systems has reduced reliance on human operators for time-critical decisions. Modern satellites can often detect potential collisions and execute avoidance maneuvers with minimal human intervention, potentially reducing collision frequency despite the increasing population of objects.

Trackability Features: Newer satellites increasingly incorporate features to enhance tracking, such as radar reflectors, optical retrore-

flectors, or transponders that continue operating beyond the satellite's primary mission life. These improvements help maintain awareness of objects' positions even after they become defunct.

The overall technological trend has been toward more debris-conscious design, though economic pressures and performance requirements sometimes work against the most sustainable options. Regulatory requirements and industry best practices have played important roles in encouraging the adoption of debris-reducing technologies.

National and Cultural Differences in Debris Generation

Analysis of historical debris data reveals distinct patterns among different spacefaring nations and organizations:

Soviet/Russian Contributions: The Soviet/Russian space program historically generated significant debris through:

The practice of intentionally separating multiple components during normal operations

Nuclear-powered satellite operations (ROSATs) that released sodium-potassium coolant droplets

Higher rates of upper stage explosions, particularly from Proton launch vehicles

However, Russian debris mitigation practices have improved significantly since the 1990s.

U.S. Patterns: U.S. debris generation has been characterized by:

Generally lower rates of mission-related debris than Soviet/Russian missions

Earlier adoption of passivation practices for upper stages

Significant contributions from Delta rocket body explosions in the 1970s-1990s

Leadership in debris tracking and research, but also in military debris-generating activities such as ASAT tests

Chinese Space Activities: China's space program has exhibited:

Rapid growth in launch rates and consequently in contribution to the debris environment

The largest single debris-generating event (2007 ASAT test)

Improving but still inconsistent compliance with international debris mitigation guidelines

European Approach: The European space program has demonstrated:

Strong emphasis on debris mitigation research and technologies

Early adoption of design-for-demise approaches

Leadership in developing international debris mitigation guidelines

Occasional significant debris events, such as the 1986 explosion of an Ariane third stage

Commercial Sector Evolution: Private space companies show varying approaches:

Established companies generally adhere to national regulatory requirements

NewSpace companies have introduced both innovative debris mitigation technologies and new challenges through large constellations

Commercial pressures sometimes create incentives to minimize mass and cost dedicated to debris mitigation features

These national and organizational differences reflect varying technical approaches, regulatory environments, and cultural attitudes toward environmental responsibility in space. Understanding these differences helps inform international policy development and targeted engagement strategies for global debris mitigation efforts.

The Role of Space Weather in Debris Population Dynamics

Solar and geomagnetic activity have played significant but often overlooked roles in the evolution of the debris environment:

Atmospheric Drag Variations: Solar activity heats and expands Earth's upper atmosphere, increasing drag on objects in low Earth orbit. During solar maxima, this natural cleaning mechanism accelerates the reentry of debris at lower altitudes. Conversely, during solar minima, debris remains in orbit longer. For example:

The extended solar minimum of 2008-2010 reduced the natural cleanup rate for LEO debris

The relatively weak solar maximum of Solar Cycle 24 (2014-2015) provided less atmospheric cleansing than previous cycles

Predictions of a potentially weak Solar Cycle 25 raise concerns about reduced natural debris removal in the 2020s

Solar Array Degradation: Solar radiation damages satellite components, particularly solar arrays, potentially leading to premature satellite failure and creating new debris objects. Variability in solar radiation thus affects satellite lifetimes and replacement rates.

Geomagnetic Storm Effects: Major geomagnetic storms can temporarily increase atmospheric density at orbital altitudes, causing un-

expected orbit changes and potentially contributing to reentry events. The March 1989 geomagnetic storm, for example, noticeably altered the orbits of thousands of tracked objects.

Long-term Climate Effects: Climate change in the lower atmosphere can affect temperatures and densities in the upper atmosphere, with some models suggesting that increased carbon dioxide levels may lead to cooling and contraction of the upper atmosphere, potentially reducing natural debris removal rates in the coming decades.

These space weather factors add complexity to debris environment modeling and highlight the interconnectedness of near-Earth space with terrestrial and solar phenomena.

Historical Debris Impacts and Their Consequences

While major collisions like the Iridium-Cosmos event are rare, numerous smaller impact events have occurred that illustrate the practical consequences of the debris environment:

Space Shuttle Window Impacts: Throughout its program history, the Space Shuttle experienced numerous impacts on its windows from small debris and micrometeoroids. NASA replaced at least 80 shuttle windows due to impact damage at a cost of approximately $50,000 per window.

Solar Panel Damages: Several satellites have reported degraded power production due to debris impacts on solar arrays. The most well-documented case is the European Space Agency's Sentinel-1A satellite, which in 2016 suffered a strike from a particle estimated to be only a few millimeters in size that nonetheless reduced power output and slightly altered the satellite's orbit and attitude.

Hubble Space Telescope Impacts: During servicing missions, astronauts observed numerous impacts on the telescope's exterior, includ-

ing a notable antenna strike. Despite these impacts, the Hubble has continued operations thanks to its robust design and the ability to perform repairs—a capability most satellites lack.

Chapter 4: Current Mitigation Methods

Technological and Operational Strategies for Mitigating Space Junk

As awareness of the space debris problem has grown, a variety of technological and operational approaches have been developed to mitigate the generation of new debris and manage the risks posed by existing debris. These strategies range from spacecraft design modifications to operational best practices and collision avoidance procedures.

Spacecraft Design for Debris Mitigation

Modern spacecraft incorporate numerous design features specifically intended to reduce their contribution to the debris environment:

Passivation Systems: To prevent explosions of defunct satellites or rocket stages, spacecraft now commonly include:

Propellant depletion burns or venting systems that release residual fuel at end-of-life

Battery discharge circuits that prevent battery explosions due to overcharging

Pressurant release valves that eliminate the risk of pressure vessel rupture These passivation features address what historically has been the largest source of fragmentation debris.

Tethered Components: Rather than releasing small components like lens covers or deployment mechanisms, modern designs often keep these items attached to the main spacecraft with tethers or hinges, preventing them from becoming separate debris objects.

Debris Shields: Particularly for critical or long-duration missions, spacecraft may incorporate Whipple shields or similar protective barriers designed to absorb the impact of small debris particles without damage to critical systems. The International Space Station uses multiple shield layers to protect habitable modules.

Self-Disposal Mechanisms: Forward-thinking spacecraft designs now routinely include:

Dedicated deorbit propulsion systems sized appropriately for end-of-mission maneuvers

Drag enhancement devices such as inflatable balloons or deployable sails to accelerate atmospheric reentry

Electrodynamic tethers that can generate drag forces through interaction with Earth's magnetic field

Design for Demise: To minimize the risk of debris surviving reentry and reaching Earth's surface, spacecraft can be designed with:

Materials that completely ablate during reentry (aluminum rather than titanium or stainless steel)

Break-up points that ensure the satellite disintegrates into smaller pieces during reentry

Components arranged to maximize heating during reentry

Collision Avoidance Capabilities: Modern spacecraft increasingly incorporate:

High-precision onboard GPS receivers for accurate position knowledge

Propulsion systems sized to accommodate collision avoidance maneuvers throughout operational life

Autonomous decision-making systems for time-critical collision avoidance

Enhanced Trackability Features: To improve space situational awareness, newer spacecraft often include:

Radar corner reflectors to enhance radar cross-section

Retroreflectors for laser ranging

Radio beacons that continue functioning beyond the primary mission lifetime

These design features, when properly implemented, can dramatically reduce a spacecraft's contribution to the debris environment. For example, a study by the European Space Agency found that implementing comprehensive passivation measures reduced the probability of explosion for a typical satellite by more than 90%.

Launch Vehicle Considerations

Launch vehicles contribute significantly to the debris environment, but several strategies can minimize this impact:

Upper Stage Disposal: Modern launch vehicles typically include:

Reignitable engines that allow controlled deorbit of upper stages

Passivation systems to prevent explosions if stages remain in orbit

Flight profiles designed to minimize orbital lifetime of any components that cannot be immediately deorbited

Mission Planning for Debris Minimization: Launch providers increasingly consider:

Orbiting in the lowest practical altitude for mission requirements

Using direct insertion techniques that reduce the number of upper stage burns

Timing launches to avoid conjunction risks with large debris objects or the International Space Station

Reusable Launch Vehicles: Systems like SpaceX's Falcon 9 with recoverable first stages inherently reduce debris risk, as these large components return to Earth rather than potentially becoming orbital debris.

Deployment Strategies: For launches carrying multiple satellites:

Careful sequencing of deployments to avoid satellite collisions

Use of dedicated deployment structures that remain attached to the upper stage

Coordination of deployment timings and directions to ensure safe separation

The implementation of these practices varies significantly among launch providers and countries. For example, ESA's Ariane 5 rocket achieved a 100% success rate in passivating its upper stage from 2014-2021, while some other providers have had lower compliance rates.

Operational Best Practices

Daily spacecraft operations offer numerous opportunities to mitigate debris risks:

End-of-Life Planning: Responsible operators develop detailed end-of-life plans that include:

Reserving sufficient propellant for disposal maneuvers

Moving to designated graveyard orbits (for GEO satellites) or lowering perigee to ensure reentry within 25 years (for LEO satellites)

Executing passivation procedures in a specific sequence to maximize safety

Orbit Selection and Management: Throughout a mission's lifetime, operators can:

Select orbits that minimize conjunction probability with existing debris clouds

Maintain precise station-keeping to stay within predetermined orbital slots

Coordinate with other operators in similar orbital regimes to maintain safe separation

Conjunction Assessment Processes: Modern satellite operations include sophisticated collision avoidance procedures:

Regular screening of predicted trajectories against the catalog of tracked objects

Risk assessment using probability calculations and covariance data

Decision-making protocols with clear thresholds for when to maneuver

Post-conjunction analysis to improve future assessments

Formation Flying and Constellation Management: For multi-satellite systems:

Automated station-keeping to maintain safe separation distances

Coordinated maneuver planning to prevent intra-constellation collisions

"Slot sharing" approaches that allow multiple satellites to safely occupy similar orbital regions

Anomaly Resolution: When spacecraft experience anomalies, debris mitigation becomes a key consideration:

Rapid safing procedures to prevent explosions or further fragmentation

Prioritization of maneuvers to move malfunctioning spacecraft out of heavily utilized orbits

Attempts to restore control specifically to enable proper end-of-life disposal

Companies and agencies with mature space debris mitigation practices, such as Intelsat, SES, and NASA, typically document these operational procedures in formal debris mitigation plans that are regularly reviewed and updated based on lessons learned and evolving best practices.

Collision Avoidance Systems and Procedures

As the orbital population increases, collision avoidance has become an essential aspect of spacecraft operations:

Conjunction Data Messaging: The standardized Conjunction Data Message (CDM) format allows for efficient communication of collision risks:

Provides details on close approaches including time of closest approach, miss distance, and probability of collision

Includes covariance data describing the uncertainty in position predictions

Enables automated processing and risk assessment

Collision Probability Calculation: Modern conjunction assessment uses sophisticated probability methods:

Monte Carlo simulations that account for uncertainties in both objects' positions

Consideration of combined hard-body radii (the physical sizes of both objects)

Evaluation of cumulative collision risk over multiple days or conjunction opportunities

Maneuver Decision Processes: Operators typically follow structured decision trees:

Low risk (typically < 1e-5 collision probability): Monitor only

Medium risk (typically 1e-5 to 1e-4): Prepare potential maneuver options

High risk (typically >1e-4): Execute avoidance maneuver unless compelling reason not to

Special Cases: Human Spaceflight: For crewed vehicles like the International Space Station:

Lower risk thresholds for avoidance maneuvers (typically 1e-4 vs. 1e-5 for uncrewed satellites)

"Shelter-in-place" procedures when warning time is insufficient for avoidance maneuvers

Dedicated tracking of debris clouds from major fragmentation events

Autonomous Collision Avoidance: Newer satellites incorporate:

Onboard processing of conjunction warnings

Self-determined maneuver planning and execution

Post-maneuver reporting to ground systems

For example, SpaceX's Starlink satellites perform autonomous collision avoidance, automatically planning and executing maneuvers without human intervention when collision risks exceed predetermined thresholds. This capability will be increasingly important as satellite populations grow and human operators cannot feasibly review every potential conjunction.

Shielding and Survivability Enhancements

For satellites that must operate in debris-dense regions, various protective measures can be implemented:

Whipple Shield Design: Named after Fred Whipple who proposed the concept in 1947, this shielding approach uses multiple layers:

An outer "sacrificial" layer that breaks up incoming particles

A standoff gap that allows fragments to disperse

An inner "catcher" layer that stops the dispersed fragments The International Space Station uses Whipple shields with multiple layers of aluminum, Kevlar, and Nextel ceramic fiber that can protect against particles up to 1-2 cm in size.

Critical Component Protection: Even when full shielding is impractical, critical systems can be hardened:

Redundant components positioned on different sides of the spacecraft

Fuel tanks with self-sealing materials or multiple containment layers

Wiring harnesses routed to minimize common-mode failure risks from penetration

Orientation Strategies: Spacecraft can be oriented to minimize vulnerability:

Presenting minimum cross-section in the velocity direction

Positioning non-critical components as "sacrificial shields" for critical systems

Rotating solar arrays edge-on to the primary debris flux direction during high-risk periods

Self-Healing Systems: Emerging technologies include:

Multilayer thermal blankets with self-sealing capabilities for micrometeoroid punctures

Pressurized systems with leak detection and isolation capabilities

Structural materials with embedded healing agents that can seal small punctures

While shielding adds mass and cost to spacecraft, its implementation represents a rational response to the increasing debris risk, particularly for high-value or long-duration missions.

Active Debris Removal Technologies

While mitigation strategies focus on preventing the generation of new debris, active debris removal (ADR) technologies aim to remove existing debris objects from orbit. This emerging field encompasses a variety of innovative approaches to capture, deorbit, or otherwise dispose of defunct satellites, rocket bodies, and fragments.

Capture Mechanisms

Successfully capturing an uncooperative object in orbit presents significant technical challenges. Several approaches have been developed or proposed:

Robotic Arms: Mechanical arms with specialized end effectors can grapple debris objects:

Advantages: Precise control, adaptable to various target shapes

Challenges: Requires close approach, sensitive to target rotation rates

Examples: ESA's e.Deorbit mission concept featured a robotic arm designed to capture Envisat, a large defunct satellite.

Nets: Deployable nets can envelop and secure irregularly shaped objects:

Advantages: Can capture tumbling objects, tolerant of position errors

Challenges: Limited control after capture, risks creating secondary debris

Examples: The RemoveDEBRIS mission successfully demonstrated net capture technology in orbit in 2018.

Harpoons: Projectiles attached to tethers can penetrate and secure to debris objects:

Advantages: Allows capture at a distance, potentially faster than other methods

Challenges: Risk of creating fragments during penetration, limited to certain material types

Examples: RemoveDEBRIS also demonstrated harpoon technology, successfully penetrating a target panel.

Grippers: Specialized mechanical devices designed to attach to specific features on target objects:

Advantages: Secure attachment, designed for specific interfaces

Challenges: Limited versatility, requires precise positioning

Examples: Astroscale's ELSA-d mission demonstrated magnetic capture technology in 2021.

Tentacles: Flexible, adaptive capturing systems:

Advantages: Can conform to irregular shapes, potentially gentler than rigid systems

Challenges: Complex control systems, lower force application capability

Examples: ESA has tested tentacle-like capture mechanisms in parabolic flight experiments.

Gecko Adhesives: Bio-inspired adhesive surfaces:

Advantages: No mechanical moving parts, potentially reusable

Challenges: Sensitive to surface conditions, limited force capacity

Examples: Stanford and JPL have tested gecko-inspired adhesives on the ISS for potential debris capture applications.

Electrostatic Attraction: Using charged surfaces to attract debris:

Advantages: Non-contact capture capability, useful for small fragments

Challenges: Limited force, affected by space plasma environment

Examples: JAXA has investigated electrostatic tethers for small debris collection.

The choice of capture mechanism depends on the target characteristics (size, shape, rotation rate), mission constraints, and risk tolerance. Most practical ADR systems will likely incorporate multiple capture options to address various debris types.

Deorbiting Technologies

Once a debris object is captured, several technologies can be employed to remove it from orbit:

Conventional Propulsion: Chemical or electric thrusters can change the object's orbit:

Advantages: Well-established technology, high reliability

Challenges: Requires significant propellant mass, especially for large objects

Examples: Conventional propulsion was proposed for ESA's e.Deorbit mission and is included in ClearSpace-1.

Drag Enhancement Devices: Deployable structures that increase atmospheric drag:

Advantages: Passive operation after deployment, no propellant required

Challenges: Extended deorbit time, limited effectiveness at higher altitudes

Examples: The DebriSat-1 design incorporated a deployable sail for end-of-life deorbit.

Electrodynamic Tethers: Conductive tethers that interact with Earth's magnetic field:

Advantages: No propellant consumption, effective in low to medium orbits

Challenges: Complex deployment, vulnerability to cutting by small debris

Examples: JAXA's KITE experiment on the ISS tested electrodynamic tether technology.

Solar Sails: Large reflective surfaces using solar radiation pressure:

Advantages: No propellant required, potentially effective at high altitudes

Challenges: Large deployable structures, limited force in LEO

Examples: NASA's NanoSail-D and The Planetary Society's LightSail demonstrated the technology.

Ion Beam Shepherd: A spacecraft that directs an ion beam at debris to gradually alter its orbit:

Advantages: No physical contact required, can work with tumbling objects

Challenges: Precise pointing requirements, extended operation time

Examples: The European LEOSWEEP project studied this concept for debris removal.

Momentum Exchange: Using tethers or mechanical systems to transfer momentum:

Advantages: Potential for multiple object removal with single system

Challenges: Complex dynamics, risk of uncontrolled behavior

Examples: Various momentum exchange tether concepts have been studied by NASA and others.

Laser Ablation: Ground or space-based lasers vaporize surface material to create thrust:

Advantages: Can affect objects without physical contact

Challenges: Requires high power, precise tracking, international policy concerns

Examples: The ORION concept studied by NASA would use ground-based lasers to deorbit small debris.

Most operational ADR missions will likely combine a capture mechanism with one or more deorbiting technologies. For example, the ClearSpace-1 mission, planned for launch in 2025, will use a robotic capture system followed by conventional propulsion to deorbit a specific upper stage rocket body.

Technical Challenges in Active Debris Removal

Despite promising technological developments, ADR faces several significant technical challenges:

Rendezvous and Proximity Operations: Approaching an uncontrolled, potentially tumbling object requires:

Precise relative navigation without cooperative sensors

Trajectory planning to safely approach tumbling objects

Contingency procedures for unexpected target behavior

Target Characterization Limitations: Limited information about debris objects complicates capture planning:

Uncertain mass properties affect capture dynamics

Unknown structural integrity may lead to breakup during capture

Surface material properties affect gripper or harpoon effectiveness

Dynamics After Capture: Controlling the combined chaser-target system presents challenges:

Center of mass shift after capture

Potential energy transfer between rotation and translation

Structural loads during deorbit maneuvers

Multiple Object Removal: Cost-effective ADR requires addressing multiple objects per mission:

Propellant requirements for transfers between debris objects

Capture system reusability considerations

Mission timeline optimization to maximize removal efficiency

Verification of Success: Confirming successful deorbit requires:

Tracking capability to verify reentry

Sufficient instrumentation to detect partial breakups

Long-term monitoring for high-altitude disposal orbits

These challenges have led to a progressive development approach, with early demonstrations focusing on cooperative targets (e.g., Astroscale's ELSA-d mission) before advancing to fully uncooperative objects.

Notable Active Debris Removal Missions and Demonstrations

Several missions have demonstrated or will demonstrate key ADR technologies:

RemoveDEBRIS (2018-2019): This Surrey Space Centre-led mission successfully demonstrated:

Net capture of a deployed target cubesat

Harpoon capture of a fixed target panel

Vision-based navigation for proximity operations

Deployable drag sail technology The mission provided valuable in-orbit validation of multiple capture technologies.

Astroscale ELSA-d (2021-2022): This commercial demonstration mission:

Deployed a small "client" satellite with a ferromagnetic docking plate

Successfully captured and released the client multiple times using magnetic capture

Demonstrated relative navigation, approach, and capture capabilities

Tested failure detection and recovery procedures

ClearSpace-1 (Planned 2025): This ESA-procured mission will:

Target a specific debris object: a VESPA upper stage adapter

Demonstrate capture of a fully uncooperative target

Perform controlled deorbit of the combined spacecraft This will be the first mission to remove an existing debris object not specifically designed for retrieval.

JAXA Commercial Removal of Debris Demonstration (CRD2) (Planned): This Japanese mission will:

Demonstrate proximity operations around a Japanese upper stage

Test technologies for future debris removal applications

Involve commercial companies in a public-private partnership

Astroscale ELSA-M (Planned):

This evolution of Astroscale's technology will:

Demonstrate the ability to remove multiple satellites from a constellation

Target satellites specifically designed with docking fixtures

Validate a business model for routine end-of-life satellite removal

DARPA RSGS (Robotic Servicing of Geosynchronous Satellites) (Planned): While primarily focused on satellite servicing, this program will demonstrate technologies directly applicable to debris removal:

Autonomous rendezvous and proximity operations

Robotic manipulation in the space environment

Long-duration operations with multiple client spacecraft

These missions represent a progression from simple technology demonstrations toward operational debris removal capabilities, with increasing complexity in terms of target characteristics and mission objectives.

Economic and Business Models for Active Debris Removal

Developing sustainable business models for ADR has proven challenging but is essential for scaling beyond government-funded demonstration missions:

Insurance-Based Models: Under this approach:

Satellite insurance providers would fund debris removal to reduce collision risks to insured assets

Premiums might be adjusted based on operators' contributions to debris removal efforts

Challenges include quantifying risk reduction and distributing costs equitably

Polluter-Pays Models: This framework would require:

Operators to post bonds or purchase "end-of-life insurance" at launch

Funds to be released upon successful self-disposal or used to contract removal services

Legal mechanisms to assign responsibility for historical debris

Government Service Contracts: Similar to the ClearSpace-1 model:

Space agencies contract private companies for debris removal services

Targets are selected based on risk assessment and national responsibility

Funding comes from public space budgets as an environmental protection measure

International Consortium Approach: This would involve:

Multiple nations pooling resources for debris removal operations

Cost-sharing based on orbital usage, historical contributions to debris, or other metrics

Collaborative selection of high-priority targets

Dual-Use Business Models: These leverage debris removal technology for multiple purposes:

Combined life extension and removal services for aging satellites

Salvage operations that recover valuable components (like solar arrays or antennas)

Using the same systems for satellite repositioning and debris removal

Regulatory-Driven Markets: Government regulations could create market demand:

Requirements for operators to remove equivalent mass to what they launch

"Cap and trade" systems for orbital debris similar to carbon markets

Launch licenses contingent on debris remediation contributions

Organizations like Astroscale are pioneering commercial approaches to debris removal, while the European Space Agency's adoption of a service-procurement model for ClearSpace-1 represents an important step toward establishing commercial viability for ADR services.

Technical Readiness and Scale-up Challenges

While individual ADR technologies have been demonstrated, scaling up to routine operations faces several challenges:

Cost Reduction Requirements: Current estimates for removing a single large debris object range from $10-150 million, depending on orbit and object characteristics. For ADR to become routine, costs must decrease substantially through:

Standardized interfaces and procedures

Reusable removal systems capable of addressing multiple objects

Economies of scale in manufacturing and operations

Reliability Considerations: Unlike many space missions, ADR involves intentional interaction with uncontrolled objects, creating unique reliability challenges:

Need for robust failure detection and recovery

Designing for wide ranges of potential target conditions

Preventing the creation of additional debris during failed removal attempts

Technology Gaps: Several key technologies require further development:

Long-duration autonomous operation systems

Standardized capture interfaces for future satellites

Non-cooperative object identification and characterization

More efficient propulsion for orbital transfers between multiple targets

Scaling Limitations: Current ADR approaches are primarily designed for removing large, intact objects one at a time. However, the debris environment contains millions of smaller objects that cannot be cost-effectively addressed with current technologies. Alternative approaches for small debris removal include:

"Debris sweepers" that can collect multiple small objects

Laser systems to alter the orbits of numerous small fragments

Large-area collectors that passively capture particles through impact

Transition from Demonstration to Operation: Moving from one-off demonstrations to routine operations requires:

Sustainable funding mechanisms

Streamlined licensing and approval processes

Development of standard operating procedures

Training pipelines for specialized personnel

Industry analysts estimate that addressing the highest-risk large debris objects in LEO would require removing approximately 5-10 objects per year over the next decade. Given current technology readiness levels, this appears technically feasible but economically challenging without new funding models or regulatory requirements.

Space Traffic Management Systems

As the number of active satellites and debris objects increases, space traffic management (STM) systems are becoming essential to prevent collisions and ensure sustainable use of orbital resources. These sys-

tems encompass the technical, regulatory, and operational frameworks needed to coordinate space activities.

Space Situational Awareness Infrastructure

Effective space traffic management begins with accurate tracking and characterization of objects in orbit:

Ground-Based Radar Systems: These systems form the backbone of current tracking capabilities:

Phased array radars like the US Space Fence can track objects as small as 5-10 cm in LEO

Mechanical tracking radars provide more detailed characterization of specific objects

Global networks are required for comprehensive coverage of all orbital regimes Examples include the US Space Surveillance Network, the German TIRA system, and Russian Okno complex.

Optical Telescopes: These complement radar systems, especially for higher orbits:

More effective than radar for tracking objects in GEO

Limited by weather, daylight, and viewing geometry

Increasingly organized into automated networks Examples include the Ground-based Electro-Optical Deep Space Surveillance (GEODSS) system and commercial networks like the ExoAnalytic Global Telescope Network.

Space-Based Surveillance: Sensors in orbit offer unique advantages:

Not limited by weather or atmospheric distortion

Can observe objects against dark space background

Positioned for optimal viewing geometry Examples include the Space Based Space Surveillance (SBSS) satellite and various experimental systems.

Radio Frequency Monitoring: Tracking satellites through their transmissions:

Can provide identity confirmation and operational status

Works even for objects too small for optical/radar tracking

Limited to active, transmitting satellites Commercial providers like Spire Global incorporate RF monitoring into their tracking services.

Data Fusion Centers: These facilities combine data from multiple sources:

Integrate government, commercial, and academic sensors

Apply advanced processing algorithms to improve orbit determination

Generate standardized products for various users Examples include the US Combined Space Operations Center (CSpOC) and the EU Space Surveillance and Tracking (SST) system.

Commercial SSA Providers: Private companies now offer tracking and conjunction analysis:

LeoLabs specializes in radar tracking of LEO objects

ExoAnalytic Solutions provides optical tracking focused on GEO

COMSPOC (Commercial Space Operations Center) offers comprehensive analysis services These commercial capabilities supplement government systems and often provide faster, more tailored services to satellite operators.

The development and integration of these systems represent significant investments. For example, the US Space Fence radar system cost approximately $1.5 billion to develop and deploy, while commercial companies have invested hundreds of millions of dollars in global sensor networks.

Catalog Maintenance and Orbit Determination

Translating raw tracking data into actionable information requires sophisticated processing:

Initial Orbit Determination: When new objects are detected:

Multiple observations are correlated to establish an initial orbit

The object is compared against the existing catalog to determine if it's new or known

Preliminary orbital elements are calculated

Orbit Refinement and Maintenance: For cataloged objects:

Regular observations are incorporated using sequential filters (often Kalman filters)

Perturbation forces are modeled (atmospheric drag, solar radiation pressure, etc.)

Maneuvers are detected and accounted for in the orbit solution

Covariance Generation: To characterize position uncertainty:

Error sources in measurements are propagated through the orbit determination process

Resulting covariance matrices describe the probability distribution of the object's position

These uncertainties are critical inputs for conjunction assessment

Special Perturbation Catalogs: Modern catalogs use high-fidelity numerical integration techniques rather than simplified analytical methods:

Incorporate detailed force models including Earth's gravitational field irregularities

Account for atmospheric density variations based on space weather conditions

Provide more accurate predictions for conjunction assessment

Require significant computational resources to maintain for large numbers of objects

The U.S. Space Surveillance Network currently tracks approximately 27,000 objects, while commercial providers track many thousands more. The processing of this tracking data represents a major computational challenge, with orbit determination processes running continuously to maintain catalog accuracy.

Conjunction Assessment and Collision Avoidance

With accurate tracking and orbit determination in place, the next critical function is predicting and avoiding potential collisions:

Screening Process: The initial identification of close approaches:

Orbits of all cataloged objects are compared to identify potential close approaches

Simplified filtering techniques identify conjunction candidates

More detailed analysis is performed on higher-risk cases

For active satellites, operators are notified when risk thresholds are exceeded

Risk Assessment: Evaluating the probability and consequences of a collision:

Probability of collision is calculated using position uncertainties from both objects

Miss distance and relative velocity determine potential impact energy

Object size, mass, and composition influence consequence assessment

Risk metrics combine probability and consequence for decision support

Collision Avoidance Maneuvers: When risk exceeds acceptable thresholds:

Satellite operators plan propulsive maneuvers to increase separation at closest approach

Maneuver timing and magnitude are optimized to minimize propellant consumption

Post-maneuver tracking confirms the new orbit and reduced collision risk

International coordination may be required when multiple satellites are involved

Operational Challenges: Practical difficulties in collision avoidance:

Limited reaction time, sometimes less than 72 hours from notification to closest approach

Uncertainties in tracking data and orbit predictions

Propellant constraints for operational satellites

Coordination difficulties when multiple operators are involved

Impossible to maneuver defunct satellites or debris objects

In 2023 alone, commercial satellite operators performed over 500 collision avoidance maneuvers, while government and military operators conducted several hundred more. This operational burden is expected to increase as both satellite populations and tracking capabilities grow.

International Guidelines and Treaties

The sustainable use of space requires not just technical solutions but also effective governance frameworks:

Current International Framework

UN Committee on the Peaceful Uses of Outer Space (COPUOS): The primary international forum:

Established in 1959 to govern the exploration and use of space for peaceful purposes

Develops principles and treaties related to space activities

Works through consensus, which can slow progress but ensures broad acceptance

Has established multiple working groups on space debris and sustainability

Outer Space Treaty of 1967: The foundation of international space law:

Establishes space as a domain to be used for peaceful purposes

Declares states responsible for national activities in space

Provides broad principles but lacks specific debris mitigation requirements

Ratified by over 100 countries but predates modern debris concerns

Registration Convention of 1976: Establishes registration requirements:

Requires states to maintain registries of objects launched into space

Mandates information sharing through the United Nations

Helps establish ownership and responsibility for space objects

Important for liability determination when debris causes damage

Liability Convention of 1972: Addresses damage caused by space objects:

Establishes absolute liability for damage caused on Earth

Fault-based liability for damage in space

Provides legal framework for compensation claims

Has rarely been invoked for debris-related incidents

IADC Space Debris Mitigation Guidelines: Technical recommendations developed by space agencies:

Limiting debris released during normal operations

Minimizing potential for break-ups during operational phases

Post-mission disposal of satellites within 25 years

Prevention of on-orbit collisions through design and operations

These guidelines form the technical basis for many national regulations

UN Space Debris Mitigation Guidelines: UN-endorsed best practices:

Adapted from IADC guidelines and endorsed by UN General Assembly in 2007

Non-binding but widely accepted as international norms

Provide common framework for national regulatory development

Complement the foundational treaties with specific debris considerations

National Regulatory Frameworks

United States: Among the most comprehensive regulatory systems:

Federal Communications Commission (FCC) requirements for commercial satellites

NASA Standard 8719.14 for government missions

Orbital debris mitigation plans required for launch licenses

Recent updates strengthening post-mission disposal requirements

Ongoing development of space traffic management framework

European Space Agency: Multi-national approach:

ESA Space Debris Mitigation Requirements

Zero Debris Charter initiative launched in 2023

Coordination across multiple European national space agencies

Development of technologies for debris prevention and removal

Japan: Advanced regulatory framework:

JAXA Space Debris Mitigation Standard

Space Activities Act of 2016 with debris provisions

Active research and development in debris mitigation technologies

China: Growing regulatory system:

National space debris prevention and control measures

Five-year plans with increasing focus on space sustainability

Testing of technologies for space debris monitoring and mitigation

Emerging Spacefaring Nations: Developing capabilities and regulations:

Often adopt international guidelines directly into national frameworks

Frequently face resource constraints in implementation

Important for comprehensive global approach to debris mitigation

Implementation Challenges

Despite the existence of guidelines and regulations, several challenges remain:

Compliance Monitoring: Difficulty verifying adherence to guidelines:

Limited transparency in some national space programs

Technical challenges in verifying compliance through observation

Asymmetric implementation across different countries and operators

Few consequences for non-compliance under current frameworks

Economic Disincentives: Cost barriers to full implementation:

Debris mitigation measures add mass, complexity, and cost to missions

Market competition can discourage voluntary measures beyond requirements

Difficulty monetizing benefits of space sustainability

Uneven regulatory requirements create competitive disadvantages

Technical Feasibility: Practical limitations in implementation:

Small satellites often lack propulsion for deorbit maneuvers

Legacy systems designed before current guidelines were established

Technical failures can prevent execution of disposal plans

Verification challenges for complex technical requirements

Emerging Commercial Activities: New challenges from novel operations:

Mega-constellations introducing unprecedented scale

On-orbit servicing changing operational paradigms

Space tourism creating new stakeholders and considerations

In-space manufacturing potentially generating new debris sources

Emerging Governance Approaches

Recognizing the limitations of current frameworks, new governance approaches are developing:

Long-term Sustainability Guidelines: Expanded framework under UN COPUOS:

Approved in 2019 after years of negotiation

21 guidelines addressing policy, regulatory, operational, and technical aspects

Broader scope than previous debris mitigation guidelines

Implementation and assessment mechanisms still developing

Space Sustainability Rating: Market-based incentive system:

Developed by World Economic Forum and partners

Rates missions based on sustainability measures

Creates transparency for investors and customers

Provides positive incentives for sustainable practices

Insurance Requirements: Financial mechanisms for responsibility:

Increasing requirements for on-orbit liability insurance

Risk-based premiums incentivizing debris mitigation measures

Potential for expanded requirements as market matures

Challenges in quantifying long-term orbital debris risks

International Coordination Mechanisms: Enhancing operator cooperation:

Space Data Association facilitating information sharing among operators

International conjunction assessment processes

Voluntary operational coordination among major satellite operators

Development of common technical standards and protocols

The evolution of these governance frameworks will be crucial to addressing the growing space debris challenge, particularly as new actors enter the space domain and novel activities emerge. While technical solutions are necessary, their implementation depends on effective governance systems that balance innovation with sustainability.

Active Debris Removal Technologies

While the preventive measures described above are essential, they cannot address the existing population of debris. Active Debris Removal (ADR) technologies aim to physically remove high-risk objects from orbit:

Capture Mechanisms

Robotic Arms: Mechanical capture systems:

Articulated manipulators adapted from terrestrial robotics

Require precise relative navigation and control

Can handle various target geometries and conditions

Examples include ESA's e.Deorbit concept and JAXA's experimental systems

Net Systems: Flexible capture solutions:

Deployable nets envelop irregular or tumbling targets

Greater tolerance for position errors than rigid systems

Successfully demonstrated in-orbit by RemoveDEBRIS mission in 2018

Challenges include net deployment dynamics and tether management

Harpoon Systems: Penetrating capture devices:

Penetrate the target structure to establish a connection

Can operate at greater distances than contact mechanisms

Also demonstrated by RemoveDEBRIS mission

Concerns about generating additional fragments during capture

Tentacle and Tether Systems: Flexible capture approaches:

Deployable tentacle-like structures to wrap around targets

Adaptable to various target geometries

Lower impact forces than harpoons or rigid mechanisms

Examples include Astroscale's ELSA-d demonstration mission

Gecko-inspired Adhesive Systems: Novel attachment technologies:

Reversible adhesive pads inspired by gecko feet

Potential for gentle, non-destructive capture

Less sensitive to target material properties

Still largely in laboratory development phase

De-orbit Technologies

Once captured, debris must be removed from crowded orbital regions:

Propulsive De-orbit: Direct removal approach:

Using the servicing spacecraft's propulsion system

Offers precise control over re-entry location

Requires significant propellant for large debris

Most straightforward approach for controlled re-entry

Drag Enhancement Devices: Passive removal method:

Deployable structures increasing atmospheric drag

Sail, balloon, or ribbon configurations

Effective in lower orbits but timeframes increase with altitude

Examples include DebriSat's drag sail and Purdue's Spinnaker3

Electrodynamic Tethers: Novel de-orbit technique:

Conductive tether interacts with Earth's magnetic field

Generates forces without propellant consumption

Challenges in deployment and dynamics control

Demonstrated in small-scale experiments like JAXA's KITE mission

Ion Beam Shepherd: Contactless removal concept:

Uses directed ion beam to transfer momentum

Avoids risks associated with physical capture

Requires precise beam control and extended operations

Studied extensively by ESA and the University of Madrid

Rendezvous and Proximity Operations

Before capture can occur, the servicing spacecraft must safely approach the target:

Relative Navigation: Finding and approaching the target:

Initial orbit determination using ground tracking

Transition to onboard sensors as range decreases

Visual, lidar, and radar systems providing relative position data

Increasing autonomy required for efficient operations

Target Characterization: Understanding the debris object:

Visual and 3D mapping of geometry

Estimation of tumble rates and stability

Material identification where possible

Assessment of structural integrity for capture planning

Trajectory Planning: Safe approach design:

Passive safety through relative orbit design

Contingency retreat paths at all mission phases

Efficiency optimization to minimize propellant use

International coordination for collision avoidance

Legal and Diplomatic Challenges: Non-technical considerations:

Permission requirements from debris owner

Liability issues during removal operations

National security concerns for certain targets

Need for transparency in operations

Mission Architecture Options

Different mission concepts address various aspects of the debris removal challenge:

Single-target Dedicated Missions: Focused removal approach:

Spacecraft designed for one specific high-risk object

Optimized capture mechanism for known target

Higher cost per object removed

Appropriate for largest, highest-risk debris

Multi-target Servicers: Economical removal approach:

Single spacecraft removing multiple objects sequentially

Requires refueling capability or highly efficient propulsion

More complex operation and mission planning

Better economics for smaller objects in similar orbits

Small-satellite Missions: Lower-cost demonstration approach:

CubeSat or microsatellite platforms for technology validation

Lower capabilities but also lower mission costs

Important for raising Technology Readiness Levels

Examples include Surrey Space Centre's RemoveDEBRIS mission

Public-Private Partnerships: Economic sustainability models:

Government funding for initial capabilities

Commercial provision of removal services

Service-based contracts rather than hardware procurement

Examples include ESA's ClearSpace-1 mission and JAXA's Commercial Removal of Debris Demonstration (CRDD)

Technology Readiness and Demonstration Missions

Several missions have demonstrated or are planning to demonstrate key ADR technologies:

RemoveDEBRIS (2018-2019): Comprehensive technology demonstration:

Successful net and harpoon capture demonstrations

Vision-based navigation experiments

Drag sail deployment test

Operated by Surrey Space Centre with European partners

ELSA-d (2021): Commercial demonstration mission:

Magnetic capture mechanism demonstration

Rendezvous operations with cooperative target

Conducted by Japanese company Astroscale

Precursor to operational debris removal services

ClearSpace-1 (Planned for 2025): First operational removal mission:

Targeting Vega Secondary Payload Adapter (vespa)

Robotic capture mechanism approach

Public-private partnership funded by ESA

Intended as first step toward commercial removal service

JAXA Commercial Removal of Debris Demonstration (Planned): Multi-phase program:

Initial removal of Japanese rocket upper stage

Testing of various capture technologies

Partnership between government and commercial entities

Focus on developing sustainable business models

Economic and Policy Considerations

Technical feasibility is only one aspect of the active debris removal challenge:

Cost Factors: Economic barriers to implementation:

Current estimates of $15-20 million per object removed

Need for dramatic cost reduction for sustainable operations

Economy of scale potential with multi-target missions

Value proposition unclear under current market conditions

Funding Models: Alternative approaches to financing:

Government-funded missions for highest risk objects

Insurance industry participation through risk reduction incentives

Launch provider contributions related to their debris creation

International cost-sharing arrangements for global benefit

Industry analysts estimate that addressing the highest-risk large debris objects in LEO would require removing approximately 5-10 objects per year over the next decade. Given current technology readiness levels, this appears technically feasible but economically challenging without new funding models or regulatory requirements.

Chapter 5: Advanced Cataloging and Orbit Refinement

5.1 Evolution of High-Precision Numerical Integration

The transition from general perturbation methods discussed in Chapter 4 to high-precision numerical integration represents a fundamental shift in space object cataloging methodology. While special perturbation catalogs laid the groundwork for systematic tracking, modern high-precision techniques have revolutionized our capacity for accurate orbit determination.

High-precision numerical integration operates by breaking down orbital motion into small time steps—typically ranging from seconds to minutes depending on the required accuracy. At each step, the complete force model is evaluated and the state vector is updated accordingly. This approach stands in stark contrast to analytical methods that rely on approximations and series expansions.

The computational backbone of this methodology involves numerical integrators such as the Runge-Kutta family of algorithms. The RK4(5) variant offers an optimal balance between accuracy and computational efficiency for most applications. For scenarios demanding extreme precision, variable step-size integrators like Dormand-Prince or Gauss-Jackson are employed, which automatically adjust integration steps based on local error estimates.

State propagation using numerical integration follows the general form:

$\ddot{r} = F(r, \dot{r}, t)/m$

Where F represents the sum of all forces acting on the object, r is the position vector, and m is the object's mass. This second-order differential equation system is typically converted to a first-order system by introducing velocity variables, resulting in six coupled equations that are solved simultaneously.

The implementation of high-precision numerical integration has enabled dramatic catalog accuracy improvements. For objects in Low Earth Orbit (LEO), position uncertainties have decreased from kilometers to meters. Similarly, for objects in Geostationary Earth Orbit (GEO), these precision gains allow for discrimination between closely spaced objects that would otherwise be indistinguishable in the catalog.

5.2 Comprehensive Force Modeling

The accuracy of orbit determination hinges critically on the fidelity of force models applied during numerical integration. Modern cataloging systems incorporate increasingly sophisticated force models to capture the complex dynamical environment experienced by orbiting objects.

5.2.1 Non-Spherical Gravity

Earth's gravitational field deviates significantly from that of a perfect sphere, primarily due to its equatorial bulge and irregular mass distribution. These deviations are mathematically expressed using spherical harmonic coefficients:

$$U(r,\varphi,\lambda) = (GM/r)[1 + \Sigma_{n=2}\wedge\infty\ \Sigma_{m=0}\wedge^{n}\ (R_0/r)^{n}\ P_{nm}(\sin\varphi)[C_{nm}\cos(m\lambda) + S_{nm}\sin(m\lambda)]]$$

Where:

U is the gravitational potential

G is the gravitational constant

M is Earth's mass

r is the distance from Earth's center

φ is latitude

λ is longitude

R_0 is Earth's equatorial radius

P_{nm} are the associated Legendre functions

C_{nm} and S_{nm} are the geopotential coefficients

Contemporary catalog maintenance routinely employs gravity models with degrees and orders up to 70×70 (EGM2008), representing over 5,000 coefficients. For objects in LEO, where non-spherical gravity effects are most pronounced, this level of fidelity is essential for maintaining accurate ephemerides over extended periods.

5.2.2 Third-Body Gravitational Effects

The gravitational pull of celestial bodies beyond Earth—primarily the Moon and Sun—introduces perturbations that significantly affect orbital motion, especially for objects in higher orbits. The third-body acceleration is computed as:

$$a_{3a} = GM_3[(r_3-r)/|r_3-r|^3 - r_3/|r_3|^3]$$

Where:

M_3 is the mass of the third body

r_3 is its position vector relative to Earth

r is the position vector of the cataloged object

These accelerations, though relatively small compared to Earth's gravity, accumulate over time and can dramatically alter orbital elements, particularly eccentricity and inclination for GEO objects.

Modern cataloging systems incorporate precise ephemerides of the Moon, Sun, and major planets from planetary theory models such as DE440, which provides submeter accuracy for lunar positions and kilometer-level accuracy for planetary positions over decades.

5.2.3 Solar Radiation Pressure

Solar radiation pressure (SRP) represents a non-gravitational force that becomes increasingly significant for objects with high area-to-mass ratios. The basic SRP acceleration model is:

$$a\_SRP = -P_0 CR(A/m)(AU/|r\_sun-r|)^2 (r\_sun-r)/|r\_sun-r|$$

Where:

P_0 is the solar radiation pressure at 1 AU (approximately 4.56×10^{-6} N/m^2)

CR is the reflectivity coefficient

A is the cross-sectional area

m is the mass

AU is the astronomical unit

r_sun is the position vector of the Sun

For high-precision catalogs, this simplified model is enhanced with:

Directional reflectivity properties using bidirectional reflectance distribution functions (BRDF)

Earth albedo and thermal radiation effects

Self-shadowing for complex spacecraft geometries

Attitude-dependent cross-sectional area variations

These refinements are particularly crucial for accurate tracking of high area-to-mass objects such as defunct satellites with deployed solar arrays or fragmentation debris consisting of lightweight materials.

5.2.4 Atmospheric Drag

For objects in LEO, atmospheric drag represents the dominant non-gravitational perturbation. The magnitude of drag acceleration is modeled as:

$$a\_drag = -\tfrac{1}{2}\rho(CD \cdot A/m)|v\_rel|v\_rel$$

Where:

ρ is atmospheric density

CD is the drag coefficient

A is the cross-sectional area

m is the mass

v_rel is the velocity relative to the atmosphere

The most significant challenge in drag modeling lies in accurately determining atmospheric density, which varies by orders of magnitude depending on solar and geomagnetic activity. Modern cataloging incorporates sophisticated atmospheric models such as NRLMSISE-00, JB2008, or DTM-2013, which account for:

Diurnal variations due to solar heating

Solar cycle effects on thermospheric density

Geomagnetic storm-induced density enhancements

Seasonal-latitudinal variations

Gravity wave effects

Recent advances include the integration of real-time atmospheric density measurements from accelerometers on scientific satellites. This allows for density corrections that significantly improve orbit prediction accuracy during periods of elevated solar activity.

5.3 Observation Integration and Catalog Maintenance

The maintenance of a high-precision catalog requires continuous integration of new observations to counteract the inevitable growth of state uncertainty over time. This process, known as catalog maintenance, involves sophisticated data assimilation techniques that optimally combine prediction models with measurement data.

5.3.1 Sequential Estimation Techniques

Modern catalogs employ sequential estimation algorithms that recursively update object states as new observations become available. The most widely implemented approach is the extended Kalman filter (EKF) and its variants:

$\hat{x}_+ = \hat{x}_- + K(z - h(\hat{x}_-))$

$P_+ = (I - KH)P_-$

Where:

$\hat{x}_+$ *is the updated state estimate*

$\hat{x}_-$ *is the predicted state*

K *is the Kalman gain matrix*

z is the observation vector

h is the observation function

P+ is the updated covariance matrix

P– is the predicted covariance

H is the linearized observation matrix

For highly non-linear orbit regimes or sparse observation scenarios, more sophisticated estimators are employed, including:

Unscented Kalman filters that better capture non-linear transformations

Particle filters that represent state probability distributions with discrete samples

Gaussian mixture models that handle multi-modal uncertainty distributions

These algorithms are particularly valuable for tracking maneuvering satellites or objects with uncertain force model parameters, as they can adaptively adjust to unexpected changes in orbital behavior.

5.3.2 Observation Weighting and Quality Control

The precision of catalog entries depends critically on proper weighting of diverse observation types. Modern systems implement sophisticated quality control mechanisms that:

Assign observation weights based on sensor-specific error models

Detect and reject outlier measurements using statistical residual analysis

Account for systematic biases in sensor networks

Adapt observation processing during conjunction events when objects are in close proximity

For radar observations, weight matrices incorporate range, azimuth, and elevation uncertainties that vary with signal-to-noise ratio and target cross-section. Optical observations are weighted based on astrometric reduction quality, atmospheric seeing conditions, and star catalog reference accuracy.

5.3.3 Regular Catalog Reprocessing

To maintain long-term catalog consistency, periodic reprocessing campaigns are conducted where historical observations are reanalyzed using the latest force models and estimation techniques. This practice ensures that:

Biases from outdated models are eliminated

Inconsistencies between objects tracked by different sensor networks are resolved

Benefits of improved computational algorithms are applied retrospectively

Object characterization parameters (area-to-mass ratios, reflection properties) are refined

Major reprocessing efforts typically coincide with significant model upgrades, such as the adoption of new Earth gravity fields or atmospheric models. These efforts can result in catalog-wide accuracy improvements of 20-40% as measured by orbit prediction performance.

5.4 Multi-Sensor Data Fusion

The evolution of space surveillance architecture has led to increasingly diverse sensor networks, necessitating sophisticated data fusion approaches to leverage their complementary strengths.

5.4.1 Sensor Network Architecture

Modern space surveillance networks integrate observations from:

Ground-based radar systems providing precise range and range-rate measurements

Optical telescopes delivering angular measurements with sub-arcsecond precision

Space-based optical sensors offering continuous coverage unaffected by weather

Laser ranging systems providing millimeter-level range measurements for cooperative targets

Radio frequency (RF) measurements including GPS-derived positions for equipped satellites

Each sensor type exhibits distinct error characteristics and sensitivity to different orbital regimes. For example, radars excel at tracking LEO objects regardless of lighting conditions, while optical telescopes are more effective for GEO objects with their slower angular motion.

5.4.2 Observation Association Algorithms

A critical challenge in multi-sensor fusion is correctly associating observations with catalog objects. State-of-the-art association algorithms employ multiple techniques:

Gating procedures that reject improbable observation-track pairs based on Mahalanobis distance

Multiple hypothesis tracking that maintains alternative association hypotheses until additional observations can resolve ambiguities

Admissible regions that constrain possible orbital solutions based on physical constraints

Machine learning methods that learn optimal association patterns from historical data

These techniques have proven particularly valuable for disambiguating dense object clusters, such as constellation satellites or fragmentation clouds, where traditional association methods struggle.

5.4.3 Sensor Tasking Optimization

To maximize catalog accuracy with finite sensor resources, advanced tasking algorithms allocate observation opportunities based on:

Current state uncertainty propagated to the observation time

Object priority based on mission importance or collision risk

Sensor capability and visibility constraints

Weather forecasts affecting optical sensors

Observation geometry quality metrics

The resulting observation schedules are dynamically updated as new measurements arrive, ensuring that resources are directed toward objects with the greatest uncertainty or operational significance.

Recent advancements include closed-loop tasking systems that automatically request follow-up observations when initial measurements indicate unexpected orbital behavior. This significantly reduces response time to potential maneuvers or breakup events.

5.5 Applications of High-Precision Catalogs

The availability of high-precision orbit catalogs enables a wide range of critical applications beyond basic space situational awareness.

5.5.1 Conjunction Analysis and Collision Avoidance

High-precision catalogs form the foundation of conjunction assessment processes that identify potential collisions between space objects. The probability of collision between two objects is computed as:

$$Pc = \iint f(x)dx$$

Where the integration is performed over the combined positional uncertainty volume, and f(x) is the probability density function of the objects' relative position.

Modern conjunction analysis systems leverage catalog covariance information to generate realistic collision probabilities that account for observation-specific uncertainties. This approach has dramatically reduced false alarms compared to simple miss-distance thresholds, allowing operators to focus resources on genuinely risky conjunctions.

For typical operational satellites, conjunction screenings are performed multiple times daily against the entire catalog, with automated processes generating alerts when probability thresholds are exceeded. This capability relies directly on the meter-level precision achieved through advanced cataloging techniques.

5.5.2 Reentry Prediction

The controlled and uncontrolled reentry of space objects represents a significant safety concern requiring precise orbit determination. High-fidelity catalog data, combined with specialized reentry models, enable predictions of:

Reentry timing with uncertainty windows as small as ±1 hour for well-tracked objects

Ground track footprints showing potential impact regions

Surviving fragment analysis for larger objects

These capabilities rely on the integration of high-precision orbital data with:

Object-specific ballistic coefficient estimation

Atmospheric density forecasts incorporating space weather predictions

Fragmentation models based on material properties and structural analysis

Monte Carlo simulations to characterize prediction uncertainties

Recent reentry prediction campaigns for high-profile objects like Tiangong-1 demonstrated the effectiveness of these techniques, with final predictions achieving accuracy within the stated uncertainty bounds.

5.5.3 Space Traffic Management

The emerging field of space traffic management leverages high-precision catalogs to develop regulatory frameworks and operational protocols for increasingly congested orbital regimes. Key applications include:

Orbit slot allocation in GEO to prevent radio frequency interference and physical conjunctions

Launch window clearance to minimize collision risks during ascent

End-of-life disposal verification to ensure compliance with orbital debris mitigation guidelines

Historical trend analysis to identify regions of increasing object density

These applications are enabled by the catalog's ability to maintain accurate knowledge of thousands of objects simultaneously, allowing for systematic analysis of orbital congestion patterns and development of evidence-based regulatory policies.

5.6 Future Directions

The evolution of space object cataloging continues to advance along several frontiers that promise further improvements in coverage, accuracy, and utility.

5.6.1 Machine Learning Integration

Machine learning techniques are increasingly augmenting traditional orbit determination methods by:

Detecting patterns in residuals that indicate unmodeled forces

Identifying object characteristics from lightcurve signatures

Predicting atmospheric density variations based on historical patterns

Optimizing observation association in ambiguous tracking scenarios

These approaches show particular promise for addressing challenges in tracking small debris and objects with complex attitude dynamics, where traditional force models often prove inadequate.

5.6.2 Collaborative Space Situational Awareness

The internationalization of space situational awareness is driving development of data sharing frameworks that integrate catalog information from multiple sources while addressing security and proprietary concerns. These collaborative approaches enable:

Improved global coverage through complementary sensor networks

Cross-validation of tracking data to identify systematic errors

Specialized tracking campaigns for high-interest events

Development of interoperable collision avoidance protocols

International data sharing initiatives have already demonstrated significant improvements in tracking accuracy and coverage compared to single-nation capabilities, particularly for objects in orbits with limited visibility from any single geographic region.

5.6.3 On-Board Orbit Determination

The miniaturization of GNSS receivers and star trackers is enabling proliferation of on-board orbit determination capabilities, even on small satellites. This development creates opportunities for:

Collaborative catalog maintenance where space objects contribute their own position data

Reduced latency in collision warnings through direct spacecraft-to-spacecraft communications

Improved autonomy for spacecraft operating beyond reliable ground control contact

Enhanced resilience against ground infrastructure disruptions

As these capabilities become more widespread, catalog architectures are evolving to incorporate self-reported position data while main-

taining robust verification mechanisms to detect spoofed or erroneous reports.

Chapter 5: Economic Implications of the Space Debris Crisis

The Hidden Costs of Orbital Debris

The space junk crisis is not merely a technical challenge—it represents a mounting economic burden that affects governments, commercial operators, and ultimately, everyday citizens. While the physical dangers of space debris are often discussed, the financial implications deserve equal consideration as they increasingly shape policy decisions and investment priorities in the space sector.

Direct Operational Costs

For satellite operators, space debris necessitates significant expenditures that were hardly considered in the early days of space exploration. These costs manifest in several ways:

Collision Avoidance Maneuvers (CAMs): Each time a satellite must adjust its orbit to avoid potential debris, operators incur costs in terms of fuel consumption, reduced operational lifespan, service interruptions, and personnel resources. The European Space Agency (ESA) alone performs hundreds of collision avoidance maneuvers annually across its fleet, with each maneuver requiring complex calculations and careful execution.

Satellite Hardening: Designing satellites with additional shielding and redundant systems to withstand potential impacts from smaller debris particles adds substantial weight and cost to spacecraft. Industry estimates suggest that hardening measures can increase satellite manufacturing costs by 5-15%, depending on the orbital environment and mission requirements.

Premature Replacements: When satellites are damaged or destroyed by debris impacts, the costs of premature replacement can be astronomical. A single communications satellite can represent an investment of $200-500 million, including manufacturing, launch, and insurance costs.

Insurance Premiums and Financial Risk

The growing debris population has not gone unnoticed by the space insurance industry. Policies for satellites in congested orbital regions now command higher premiums that reflect the increased collision risks. Insurance rates for satellites in low Earth orbit (LEO) have risen by approximately 20-30% over the past decade, primarily due to debris-related concerns.

Some orbital regions have become so hazardous that insurers have begun imposing special conditions or even excluding coverage for certain collision scenarios. This shifting insurance landscape has profound implications for business models in the space industry, particularly for newer entrants with limited financial reserves.

Economic Value at Risk

The satellite services sector generates over $300 billion in annual revenue globally, supporting critical infrastructure for communications, navigation, weather forecasting, and Earth observation. This economic activity—and the terrestrial economies that depend on it—faces increasing vulnerability as the debris population grows.

A 2023 analysis by the Organisation for Economic Co-operation and Development (OECD) estimated that a major collision cascade in LEO could disrupt satellite services worth $25-30 billion annually, with secondary economic impacts potentially reaching into the trillions of dollars when considering dependencies across sectors like transportation, agriculture, and emergency services.

Consider the following examples of economic dependency:

Global navigation satellite systems (GNSS) support approximately 6-8% of GDP in advanced economies through applications in logistics, precision agriculture, and financial services.

Weather satellites provide data that improves forecast accuracy, delivering economic benefits estimated at $160 billion annually in the United States alone.

Communications satellites facilitate transactions worth billions of dollars daily in the global financial system.

The Cleanup Conundrum: Allocating Costs

While the need for active debris removal (ADR) is increasingly apparent, the question of who should bear the costs remains contentious. Current estimates place the cost of removing a single large debris object between $10-15 million, making large-scale cleanup operations prohibitively expensive without new funding mechanisms.

Several economic approaches are under consideration:

Polluter Pays Principle: Requiring satellite operators to pay for the eventual removal of their own spacecraft, either through direct responsibility or contributions to a remediation fund.

Orbital Use Fees: Implementing a system of fees for the use of orbital slots, with proceeds directed toward debris remediation efforts.

Public-Private Partnerships: Governments subsidizing initial ADR demonstration missions while commercial entities develop more cost-effective removal technologies and services.

International Fund: Creating a globally managed fund with contributions proportional to nations' space activities and economic capabilities.

The New Space Economy: Threats and Opportunities

The debris problem presents both challenges and opportunities for the emerging commercial space sector:

Threats to Growth

The current trajectory of debris accumulation threatens to undermine the anticipated growth of the space economy, which some analysts project could reach $1-3 trillion by 2040. New constellations of thousands of satellites face particular risks, as their business models often depend on long-term operations in already congested orbital regions.

Venture capital investment in space startups—which exceeded $15 billion in 2021—could diminish if investors perceive increasing debris-related risks to business sustainability.

Economic Opportunities

Conversely, the debris challenge has spawned entirely new business sectors focused on:

Space Situational Awareness (SSA): Companies offering enhanced debris tracking and collision prediction services represent a growing market segment, with annual revenues projected to reach $1.5 billion by 2028.

End-of-Life Services: Specialized services for satellite deorbiting assistance and "space tugs" that can remove defunct satellites from valuable orbital slots.

Debris Removal Technologies: Firms developing technologies for active debris removal, from harpoons and nets to robotic arms and laser systems.

Sustainable Space Infrastructure: On-orbit servicing, assembly, and manufacturing capabilities that could reduce the need for replacement launches and associated debris generation.

The Cost of Inaction

Perhaps the most significant economic consideration is the potential cost of failing to address the debris problem. Studies by NASA, ESA, and independent researchers suggest that if current trends continue, certain orbital regions could become unusable within decades due to collision cascades.

The resulting loss of access to these orbits would represent an economic catastrophe of unprecedented scale—effectively surrendering trillions of dollars of future economic activity and forfeiting critical infrastructure upon which modern civilization increasingly depends.

A 2024 economic analysis published in the Journal of Space Economics estimated that each dollar invested in debris remediation today could prevent $7-10 in future economic losses, making debris mitigation one of the most cost-effective forms of infrastructure protection available to policymakers.

Toward Sustainable Space Economics

The economic dimensions of the space debris crisis point toward the need for new approaches that align financial incentives with sustainable practices:

Life-Cycle Pricing: Incorporating the full costs of responsible spacecraft disposal into mission budgets from the outset.

Market-Based Mechanisms: Developing tradable permit systems for orbital slots that reward operators who minimize debris generation and penalize those who create higher risks.

Economic Valuation of Orbits: Recognizing orbital paths as limited natural resources with quantifiable economic value that deserves protection.

International Economic Frameworks: Creating globally recognized systems for liability, compensation, and cost-sharing related to debris mitigation and remediation.

Conclusion

The economic implications of the space debris crisis extend far beyond the immediate costs of damaged satellites or collision avoidance maneuvers. At stake is the long-term viability of space as an economic domain—a frontier that promises tremendous benefits but requires sustainable management.

By developing economic frameworks that accurately value orbital resources, incentivize responsible behavior, and distribute cleanup costs equitably, the international community can help ensure that space remains accessible for future generations of commercial and scientific endeavors. The alternative—allowing short-term economic considerations to drive unsustainable exploitation of orbital environments—would represent not just an environmental failure but an economic catastrophe of historic proportions.

Appendix A: Major Space Debris Events

Date	*Event*	*Debris Created*	*Significance*
April 1968	*Fragmentation of Cosmos 21*	*~10 trackable pieces*	*First recorded major fragmentation event*
June 1983	*Deliberate fragmentation of NOAA-8*	*~250 trackable pieces*	*Early demonstration of anti-satellite capabilities*
September 1986	*SPOT-1/Viking rocket body explosion*	*489 trackable pieces*	*Major unintentional fragmentation*
January 2007	*Chinese anti-satellite test (Fengyun-1C)*	*>3,000 trackable pieces*	*Largest deliberate debris-creating event to date*
February 2009	*Iridium 33-Cosmos 2251 collision*	*>2,300 trackable pieces*	*First major accidental collision between intact satellites*
March 2021	*Fragmentation of Yunhai 1-02*	*>40 trackable pieces*	*Suspected collision with debris from Zenit-2 rocket*
November 2021	*Russian ASAT test (Cosmos 1408)*	*>1,500 trackable pieces*	*Created debris field threatening ISS operations*

Appendix B: Glossary of Space Debris Terminology

Active Debris Removal (ADR): Technologies and missions specifically designed to remove existing debris objects from orbit.

Breakup: The fragmentation of a space object, either through explosion, collision, or structural failure.

Collision Avoidance Maneuver (CAM): An orbital adjustment performed by a spacecraft to reduce the risk of collision with another object.

Conjunction: A close approach between two objects in orbit.

Deorbit: The process of removing a satellite from orbit, typically by allowing it to reenter Earth's atmosphere.

Graveyard Orbit: A disposal orbit above GEO where satellites are placed at the end of their operational life.

Kessler Syndrome: A theoretical scenario where the density of objects in LEO becomes high enough that collisions between objects generate a cascade of further collisions.

Orbital Decay: The gradual decrease in orbital altitude due to atmospheric drag.

Passivation: The process of depleting a satellite's energy sources (fuel, batteries) at end-of-life to prevent future explosions.

Space Situational Awareness (SSA): The comprehensive knowledge of space objects and the space environment, including tracking, characterizing, and predicting the positions of objects in orbit.

Space Traffic Management (STM): The planning, coordination, and execution of activities to enhance the safety, stability, and sustainability of operations in the space environment.

Trackable Object: A space object large enough (typically >10 cm in LEO) to be reliably tracked by current surveillance systems.

Appendix C: Key Organizations and Initiatives

International Organizations

Inter-Agency Space Debris Coordination Committee (IADC)

A forum of governmental bodies for the coordination of activities related to space debris issues.

United Nations Committee on the Peaceful Uses of Outer Space (UN-COPUOS)

The primary UN body addressing legal and technical aspects of space activities, including debris mitigation.

Space Data Association (SDA)

A non-profit organization that allows satellite operators to share data to improve space situational awareness.

National Space Agencies

NASA Orbital Debris Program Office

Conducts measurements of the orbital debris environment and develops mitigation measures.

European Space Agency Space Debris Office

Coordinates ESA's research activities in the field of space debris.

Japan Aerospace Exploration Agency (JAXA) Space Debris Office

Focuses on monitoring space debris and developing removal technologies.

Notable Private Initiatives

Astroscale

A private company developing technologies for on-orbit servicing and active debris removal.

ClearSpace

A Swiss company developing technologies for capturing and removing space debris.

LeoLabs

Provides commercial tracking services for objects in low Earth orbit.

Appendix D: International Legal Framework

Existing Treaties and Agreements

Outer Space Treaty (1967)

Establishes that states bear international responsibility for national activities in space and should avoid harmful contamination of space.

Liability Convention (1972)

Establishes that launching states are liable for damage caused by their space objects.

Registration Convention (1975)

Requires states to maintain a registry of objects launched into space and provide information to the UN.

Non-Binding Guidelines

IADC Space Debris Mitigation Guidelines (2002, updated 2007)

Technical recommendations for minimizing the creation of space debris.

UN Space Debris Mitigation Guidelines (2007)

Non-binding guidelines endorsed by the UN General Assembly.

The Hague International Code of Conduct against Ballistic Missile Proliferation (2002)

Includes provisions relevant to limiting debris from missile tests.

Emerging Frameworks

Guidelines for the Long-term Sustainability of Outer Space Activities (2019)

Adopted by UNCOPUOS, these guidelines address broader sustainability issues including debris mitigation.

The Artemis Accords (2020-present)

Bilateral agreements led by the United States that include provisions on spacecraft disposal and debris mitigation.

Appendix E: Future Research Directions

Technology Development Priorities

Large Debris Removal

Technologies for safely capturing and deorbiting large intact objects.

Small Debris Remediation

Methods for addressing the numerous small debris objects that cannot be tracked individually.

On-Orbit Servicing

Capabilities to repair, refuel, or upgrade satellites to extend their useful life and prevent them from becoming debris.

Sustainable Materials

Development of materials that minimize fragmentation or that degrade more rapidly in the space environment.

Policy Research Needs

Economic Models

Research into sustainable funding mechanisms for debris remediation.

International Governance

Studies of effective institutional arrangements for space sustainability.

Compliance Mechanisms

Investigation of methods to ensure adherence to debris mitigation guidelines.

Liability Frameworks

Development of clear standards for responsibility and compensation related to debris-causing events.

Appendix F: Resources for Further Learning

Books

Bradley, Allen J., and Lawrence D. Roberts. "Orbital Ecology: Space Environmental Management in the 21st Century," Oxford University Press, 2023.

Klinkrad, Heiner. "Space Debris: Models and Risk Analysis," Springer, 2006.

Pelton, Joseph N. "Space Debris and Other Threats from Outer Space," Springer, 2013.

Academic Journals

Journal of Space Safety Engineering

Advances in Space Research

Acta Astronautica

Space Policy

Online Resources

European Space Agency's Space Debris Office website

NASA Orbital Debris Program Office website

Space-Track.org (satellite tracking data)

The Secure World Foundation resources on space sustainability

Conferences

International Astronautical Congress (annual)

European Conference on Space Debris (every 4 years)

Advanced Maui Optical and Space Surveillance Technologies Conference (annual)

About the Author

Jonathon Wetzel is a multifaceted author exploring the intersection of artificial intelligence, human potential, and the unknown. A veteran with a deep interest in the UAP enigma and its potential links to AI advancements, he brings a unique perspective to both technological innovation and self-improvement. His experience with cutting-edge AI tools informs his practical self-help and self-development guides, while his passion for understanding the world's mysteries fuels his explorations of the extraordinary. A father of two, Jonathon strives to empower others to embrace the future, whether it involves mastering AI or unlocking their inner strength.

www.ingramcontent.com/pod-product-compliance
Lightning Source LLC
LaVergne TN
LVHW022102200926
848481LV00003B/1035
* 9 7 9 8 2 2 7 6 5 8 7 9 1 *